Challenging Programming in Python: A Problem Solving Perspective

Habib Izadkhah · Rashid Behzadidoost

Challenging Programming in Python: A Problem Solving Perspective

Springer

Habib Izadkhah
Department of Computer Science
University of Tabriz
Tabriz, Iran

Rashid Behzadidoost
Department of Computer Science
University of Tabriz
Tabriz, Iran

ISBN 978-3-031-40001-8   ISBN 978-3-031-39999-2 (eBook)
https://doi.org/10.1007/978-3-031-39999-2

© The Editor(s) (if applicable) and The Author(s), under exclusive license to Springer Nature Switzerland AG 2024

This work is subject to copyright. All rights are solely and exclusively licensed by the Publisher, whether the whole or part of the material is concerned, specifically the rights of translation, reprinting, reuse of illustrations, recitation, broadcasting, reproduction on microfilms or in any other physical way, and transmission or information storage and retrieval, electronic adaptation, computer software, or by similar or dissimilar methodology now known or hereafter developed.
The use of general descriptive names, registered names, trademarks, service marks, etc. in this publication does not imply, even in the absence of a specific statement, that such names are exempt from the relevant protective laws and regulations and therefore free for general use.
The publisher, the authors, and the editors are safe to assume that the advice and information in this book are believed to be true and accurate at the date of publication. Neither the publisher nor the authors or the editors give a warranty, expressed or implied, with respect to the material contained herein or for any errors or omissions that may have been made. The publisher remains neutral with regard to jurisdictional claims in published maps and institutional affiliations.

This Springer imprint is published by the registered company Springer Nature Switzerland AG
The registered company address is: Gewerbestrasse 11, 6330 Cham, Switzerland

Paper in this product is recyclable.

Preface

Programming is a fascinating field that requires creativity, problem-solving skills, and curiosity. Python is a popular and versatile programming language that is widely used in various domains, from data science and machine learning to web development and scientific computing. Python's clean syntax, vast ecosystem of libraries, and dynamic nature make it an ideal language for solving complex problems. Programming forces individuals to think logically and precisely, as the process of reaching the output must be formulated accurately. Therefore, it is essential to have books that provide programmers with the ability to solve challenging problems. On the other hand, books are also necessary to improve human thinking and reasoning skills to solve problems in daily life and work. Creative thinking and logical reasoning are crucial for solving problems, and this book aims to achieve two general goals: (1) improving thinking and reasoning skills by investigating and programming challenging problems and (2) enhancing Python programming skills by presenting challenging problems and solving them step by step. This book is for those who want to take their Python skills to the next level and tackle challenging programming problems. From a programming perspective, this book is beneficial for individuals with elementary, intermediate, and advanced Python programming skills and anyone who wishes to learn a programming language proficiently enough to solve complex problems. The basics of Python are instructed from the ground up in this book, with numerous examples, and challenging problems with Python codes, algorithms, and notes. We have attempted to achieve the two goals by presenting and solving 90 challenging problems from various areas. Each chapter focuses on a specific type of challenge, increasing the reader's interest in following the challenging problems further. This book is divided into eight chapters, starting with the basics of learning Python programming language in Chap. 1, followed by presenting the necessary Python basics for programming challenging problems in Chap. 2. The subsequent chapters focus on specific types of challenges, such as math-based challenges in Chap. 3, number-based challenges in Chap. 4, string-based challenges in Chap. 5, game-based challenges in Chap. 6, count-based challenges in Chap. 7, and miscellaneous challenges in Chap. 8. This book's audience includes students from all fields, ranging from zero to advanced levels, as well as teachers and individuals interested in improving their Python programming skills. Additionally, this book is useful for students planning to participate in programming competitions. After learning the topics presented in this book, the learner will be capable of coding challenging problems in Python.

Tabriz, Iran
Habib Izadkhah
Rashid Behzadidoost

Contents

1	**Introduction**		1
	1.1 Why Python?		1
	1.2 Non-library Based		2
	1.3 Enhancing Programming Skills and Creative Thinking Through Challenging Problems		2
	1.4 Prerequisite		3
	1.5 Target Audience		3
2	**Python Basics**		5
	2.1 How to Run a Python Program		5
	2.2 Data Types in Python		8
		2.2.1 Boolean Type	9
		2.2.2 Integer Type	9
		2.2.3 Decimal Type	12
		2.2.4 String Type	12
	2.3 Data Structures		28
	2.4 List		28
		2.4.1 List Built-in Functions	29
		2.4.2 List Constructor	36
	2.5 Array		36
		2.5.1 Sort	37
	2.6 Matrix		37
		2.6.1 Changing List Elements	38
	2.7 Set		39
		2.7.1 Set Built-in Functions	39
		2.7.2 Set Constructor	45
	2.8 Dictionary		46
		2.8.1 Nested Dictionary	47
		2.8.2 Dictionary Built-in Functions	48
		2.8.3 Dictionary Constructor	51

vii

2.9	Tuple		51
	2.9.1	Nested Tuple	53
	2.9.2	Tuple Built-in Functions	53
	2.9.3	Tuple Constructor	53
	2.9.4	Range Type	54
2.10	Statements		55
2.11	Conditions		58
2.12	Loops		60
	2.12.1	For	60
	2.12.2	Comprehensions	62
2.13	Functions		64
	2.13.1	Variable Scopes	68
	2.13.2	Lambda Function	69
	2.13.3	Handling the Control Flow	70
2.14	Modules		73
2.15	Generators		74
2.16	Recursion		75

3 Math   77

3.1	Josephus	78
3.2	Reaching a Point in Lattice Path	81
3.3	Brussel Choice Problem	82
3.4	Inverse Collatz Conjecture	84
3.5	Counting Possible Corners	88
3.6	Nearest S-gonal	89
3.7	Finding Fulcrum Position	90
3.8	Counting Sphere Pyramid Blocks	92
3.9	Grouping Coins	94
3.10	Median of Triple Medians	95
3.11	Smallest Seven-Zero	97
3.12	Postfix Evaluate	100
3.13	Stable State in Bulgarian Solitaire	102
3.14	Computing the Rectangular Towers in Manhattan Skyline	105
3.15	Cut Rectangular into Squares	108
3.16	Eliminating Corners	111
3.17	Leibniz Triangle	114
3.18	Collatzy Distance	117
3.19	Sum of Two Squares	120
3.20	Has Three Summer	121
3.21	Perfect Power	125
3.22	Lunar Multiplication	126
3.23	n-th Term of Recaman Sequence	129
3.24	Van Eck Sequence	131
3.25	Non-consecutive Fibonacci Numbers	133
3.26	Fibonacci Word	135

	3.27	Finding the Most Point Line	136
	3.28	Is a Balanced Centrifuge	139
4	**Number**		143
	4.1	Cyclop Numbers	144
	4.2	Is a Domino Cycle	145
	4.3	Extract Increasing Digits	146
	4.4	Expand Integer Intervals	148
	4.5	Collapse Integer Intervals	150
	4.6	Left-Handed Dice	151
	4.7	Bonus to Repeated Numbers	153
	4.8	Nearest First Smaller Number	156
	4.9	First Preceding by K Smaller Numbers	158
	4.10	n-th Term of Calkin Wilf	159
	4.11	Reverse Ascending Subarrays	162
	4.12	Smallest Integer Powers	163
	4.13	Sorting Cycles of a Graph	165
	4.14	Obtaining Numbers in Balanced Ternary System	167
	4.15	Is Strictly Ascending	169
	4.16	Priority Sorting	171
	4.17	Sorting Positives, Keep Negatives	173
	4.18	Numbers First, Characters Second	175
	4.19	Sorting Dates	177
	4.20	Sorting by Alphabetical Order and Length	178
	4.21	Sorting by Digit Count	180
5	**String**		183
	5.1	Pancake Scramble into Texts	183
	5.2	Reverse Vowels into Texts	185
	5.3	Word Shape from a Text Corpus	186
	5.4	Word Height from a Text Corpus	187
	5.5	Color Combination	191
	5.6	McCulloch Second Machine	194
	5.7	Champernowne Word	195
	5.8	Combining the Strings into One New String	197
	5.9	Unscrambling the Given Word	200
	5.10	Auto-Correcter Word	201
	5.11	Correct Verb Form in Spanish	203
	5.12	Subsequent Letters	207
	5.13	Possible Words from a Text Corpus	209
6	**Game**		213
	6.1	Winner in Card Game	213
	6.2	Hand Shape in Bridge Game	218
	6.3	Contract Bridge Game	219
	6.4	Same Hand Shape Distribution	222

	6.5	Number Round Counter	224
	6.6	Reaching Stable State in Candy Share	226
	6.7	Oware Game	228
	6.8	Safe Squares in Chessboard	230
	6.9	Safe Squares Not Threaten with Bishops	231
	6.10	Reaching Knight Jump	233
	6.11	Capturing Maximum Checkers	235
	6.12	Safe Rooks with Friends	240
	6.13	Highest Score in Crag Score	242
	6.14	Optimal Crag Score with Multiple Rolls	244
7	**Count**		249
	7.1	Counting Number of Carries	249
	7.2	Counting Growl of Animals	251
	7.3	Counting Consecutive Summers for a Polite Number	254
	7.4	Counting Occurrence of Each Digit	255
	7.5	Counting Maximal Layers	257
	7.6	Counting Dominator Numbers	258
	7.7	Counting Troikas from Integer Numbers	260
	7.8	Counting Intersected Disks	261
8	**Miscellaneous Problems**		265
	8.1	Riffling Items	265
	8.2	Calculate Smaller Coins	267
	8.3	Keep Frequent Items at Most n	269
	8.4	Collision Time of Frogs	270
	8.5	Positioning in Wythoff Array	274
	8.6	Fractran Interpreter	277

About the Authors

Dr. Habib Izadkhah is an associate professor at the Department of Computer Science, University of Tabriz, Iran. He worked in the industry for a decade as a software engineer before becoming an academic. His research interests include algorithms and graphs, software engineering, and bioinformatics. More recently, he has been working on developing and applying deep learning to a variety of problems, dealing with biomedical images, speech recognition, text understanding, and generative models. He has contributed to various research projects, authored a number of research papers in international conferences, workshops, and journals, and also has written five books, including *Source Code Modularization: Theory and Techniques* from Springer and *Deep Learning in Bioinformatics* from Elsevier.

Rashid Behzadidoost is a Ph.D. candidate in Computer Science at the University of Tabriz, Iran. He is currently pursuing his doctoral degree in Computer Science, specializing in artificial intelligence, and natural language processing. Rashid has a deep passion for coding and enjoys solving challenging problems. He has obtained his skills through years of study, practice, and teaching. He has taught several courses on computer sciences including challenging programming, microprocessor, and data structure at the University of Tabriz.

Chapter 1
Introduction

This chapter discusses the objectives, applications, and necessities of this book.

1.1 Why Python?

Python is a high-level programming language with a simple syntax compared to other programming languages. In addition, Python is a general-purpose, cross-platform, multi-paradigm, object-oriented language that supports dynamic data types. Python is easy to understand, and even individuals with a basic understanding of programming can learn Python quickly. The primary reason for this simplicity is that Python programming language statements are similar to the English language, making the learning process straightforward. Python has an interactive coding environment that makes working with the language and testing command execution while learning easy. Being cross-platform is a significant advantage of Python because it can be used on various operating systems such as Mac, Windows, Linux, and even iOS and Android. One of the most significant advantages of Python is that it offers extensive libraries for most work areas. In fact, the library provides many ready-made code pieces that programmers can use in their work. For instance, to connect Python to a database, there is no need to specify all the details of the database connection by coding. Instead, to connect to the database, one only needs to call the library's name. All of these features make Python an excellent language to learn. Furthermore, Python has numerous real-world applications, such as web development, game development, data science, and artificial intelligence. In addition to its simplicity and versatility, Python is also a popular choice among developers due to its active and supportive community. As an open-source language, Python has a vast community of developers who contribute to its growth and development. This community provides a wealth of resources, including documentation, forums, and tutorials, making it easier for new learners to get started and for experienced developers to find solutions to their problems. Additionally, this community is continually updating and improving Python through new packages, libraries, and frameworks, ensuring that the language remains relevant and up-to-date in an ever-changing technological landscape. This supportive and dynamic community is another reason why Python is an excellent language to learn.

1.2 Non-library Based

Most published Python books typically either cover the language's syntax with basic examples or focus on packages for problem-solving. However, using libraries and simple exercises to understand Python's structure can often be insufficient in providing developers with the necessary skills to tackle complex problems.

We believe that such approaches are inadequate for professional programmers who are dealing with complex programs. Therefore, this book takes a different approach by discussing problems that are not library-based (although standard libraries are used in some programs), and presents solutions through detailed step-by-step algorithms, hints, examples, and commented code. This approach is designed to equip programmers with the necessary skills to solve more complex problems, leading to an improvement in their overall programming abilities.

1.3 Enhancing Programming Skills and Creative Thinking Through Challenging Problems

Solving challenging problems is an effective way to improve programming skills and foster creative thinking. Programmers must first identify a problem, devise a solution, and then translate it into an algorithm before implementing it in a specific programming language. By tackling difficult problems, programmers learn how to approach complex issues systematically and improve their programming skills. This book aims to enhance the power of thinking and reasoning among programmers and deepen their understanding of the Python language by presenting 90 challenges that are coded step-by-step. These challenges are grouped into chapters on math, numbers, strings, games, counting, and miscellaneous topics. Each chapter contains a set of challenges with examples, hints, and Python code solutions. The number of challenges in each chapter is as follows: (1) math—28, (2) numbers—21, (3) strings—13, (4) games—14, (5) counting—8, and (6) miscellaneous—6.

1.4 Prerequisite

In addition to the chapters on challenging problems, a separate chapter on Python basics has been included to simplify the programming process for the presented challenges. This chapter provides examples and notes to teach everything necessary to understand the basics of Python and to tackle the challenges. Therefore, there are no prerequisites for studying this book.

1.5 Target Audience

This book presents and solves 90 challenges that can benefit a diverse range of people, including computer science and engineering students, as well as anyone who wants to master the Python language. The presented problems cover various fields, making it suitable for a wide range of people. Moreover, this book enhances the power of thinking and reasoning irrespective of the programming language. It can serve as a valuable resource for teaching Python in universities or schools. Additionally, software developers and participants in ACM or other competitions can benefit from this book to improve their programming skills.

Chapter 2
Python Basics

This chapter provides an overview of the fundamental concepts and knowledge of Python required for this book. It includes instructions for installing and running a simple Python program, covers the concept of variables and the basic data types of boolean, integer, float, string, and the sequence type of range. The chapter also discusses data structures such as lists, arrays, matrices, dictionaries, and sets, as well as various types of if statements with examples. It covers loop statements, including for and while loops, as well as nested loops, and provides examples for each type. Finally, the chapter provides examples for creating built-in functions, user-defined functions, lambda functions, modules, and generators. Note that this chapter does not cover all statements and modules of Python.

2.1 How to Run a Python Program

This book uses Python version 3.9 and Windows 10 operating system. An integrated development environment (IDE) is a programming environment that offers useful features for programmers such as an editor, debugger, and code completion. The use of an IDE simplifies Python coding. There are several popular IDEs available, including Pycharm, Jupyter, and Spyder. The choice of IDE depends on personal preference, as most IDEs use a Python interpreter, and their differences are mainly in appearance and features. In this book, Spyder IDE is used for Python programming, which requires the installation of Anaconda. Anaconda is an open-source distribution that enables users of Windows, Linux, and other operating systems to code in Python and R programming languages. The installation of Anaconda is a straightforward process, like any other software, requiring only a few simple clicks. Figure 2.1 depicts how the Anaconda must be installed. Once Anaconda is installed on Windows 10, Spyder can be accessed by searching for 'Spyder' in the Windows search bar until

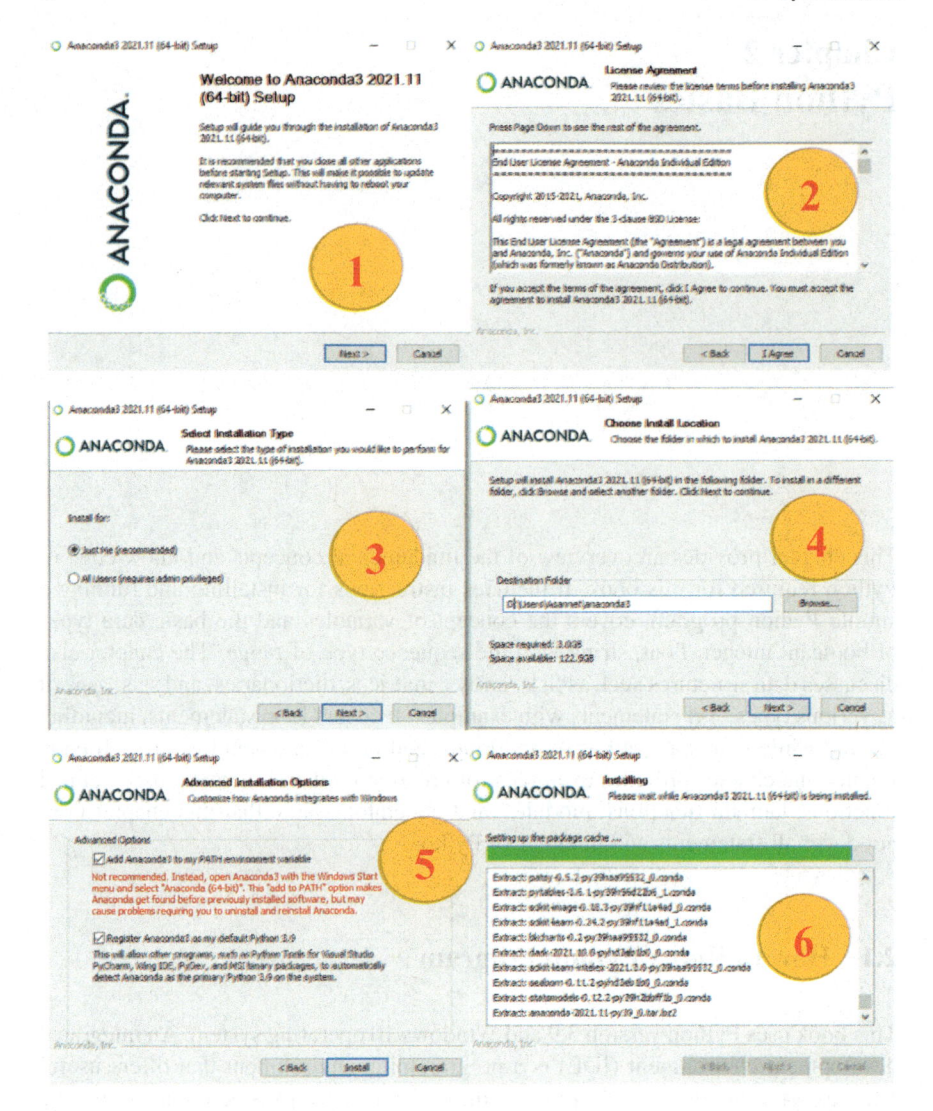

Fig. 2.1 Steps to install Anaconda

the Spyder icon is displayed. Clicking on the icon will open the Spyder environment, as depicted in Fig. 2.2.

It is important to note that Spyder requires the installation of Anaconda, as mentioned earlier in the chapter.

Figure 2.3 depicts the Spyder environment that specifies the function of each block, also, the $F5$ key can be used to run the written code.

2.1 How to Run a Python Program

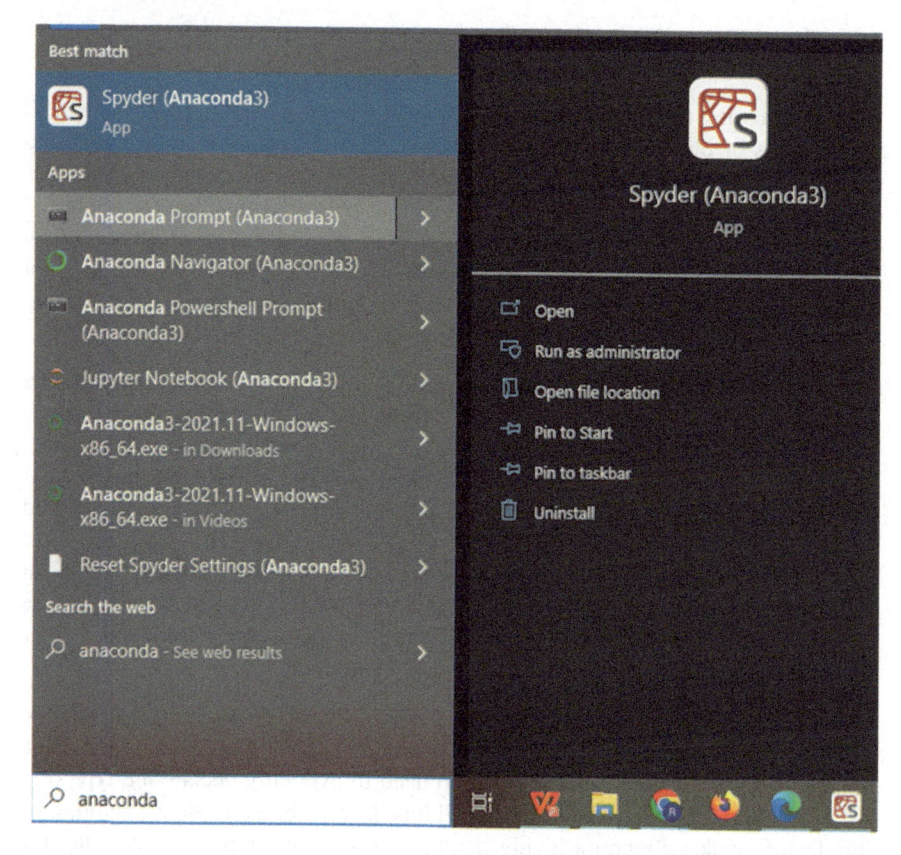

Fig. 2.2 Opening Spyder

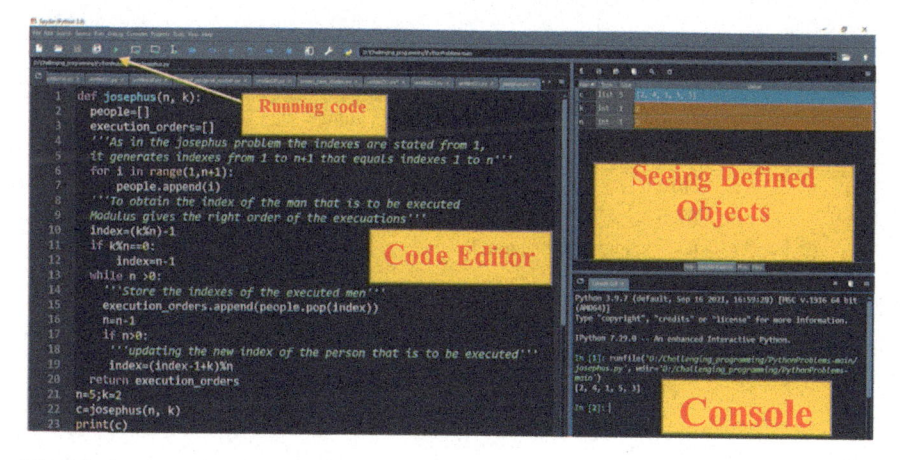

Fig. 2.3 Spyder environment

2.2 Data Types in Python

Before discussing the supported data types in Python, it is important to understand the concepts of objects and variables. Python is an object-oriented programming language, meaning that everything in Python is an object. Object orientation enables programmers to perform tasks by defining various objects with unique attributes.

In Python, a variable is essentially a word that acts as a pointer to an object. When an object is assigned to a variable, the variable is created automatically and does not need to be defined beforehand. It is imperative to follow certain rules while writing a variable in Python, such as naming conventions and avoiding reserved keywords.

1. A variable in Python must follow specific naming conventions. It should start with a letter or an underscore, and can be followed by any combination of letters, digits, and underscores. For example, '_sel' or 'tryj' are valid variable names.
2. Python is a case-sensitive language. This means that variables with different capitalization, such as 'Tryj' and 'tryj', are treated as distinct and cannot be used interchangeably.
3. Certain words in Python, known as reserved words or keywords, have special meanings within the language and cannot be used as variable names. These words are reserved for specific purposes and cannot be repurposed. Please refer to Table 2.1 for a list of reserved words in Python.

Unlike statically typed languages, where variable types are determined at compile time, in Python, the type of a variable is determined at runtime based on the value assigned to it. This means that there is no need to explicitly declare the type of a variable before using it, as the interpreter infers the type based on the value it holds. This feature makes Python a highly flexible and dynamic language, allowing for rapid development.

Table 2.1 The keywords in Python programming language

False	None	True	and	as
assert	break	class	continue	def
del	elif	else	except	finally
for	from	global	if	import
in	is	lambda	nonlocal	not
or	pass	raise	return	try
while	with	yield		

2.2.1 Boolean Type

In the real world, there are numerous objects that can exist in only two distinct states. For example, an electronic device can be switched on or off, an individual can agree or disagree with a particular stance, or a concept can have two opposing states. In Python, the 'True' keyword signifies that the state is true, while the 'False' keyword signifies that the state is false. In actuality, 'True' is equivalent to the numerical value of 1, while 'False' is equivalent to 0.

2.2.2 Integer Type

In practice, integers are the most commonly used numeric type in real-life applications. They are widely used for counting odd and even numbers, tallying monetary values, and similar tasks. For instance, the following code shows the addition of two integer values:

```
1  x=2+3
2  print(x)
3  # Output: 5
```

There are several points to note regarding the code provided. Firstly, in line 1, a variable named x is defined and assigned an integer value. Secondly, in line 2, the $print(x)$ statement is used to display the value stored within the variable x. Thirdly, line 3 is a comment. In Python, comments are used to add annotations in the source code for the purpose of improving its readability. Comments in Python are denoted by the symbol #. It is worth noting that no explicit declaration for the variable x is needed in Python. This is due to Python's dynamic-typing feature, which allows variables to be defined without specifying their type beforehand.

In the next some examples are provided in the below. Computing the remainder of two numbers with notation % is performed, as follows:

```
1  x=2%3
2  print(x)
3  # Output: 2
```

Computing the multiplication of two numbers with notation * is performed, as follows:

```
1  x=2*3
2  print(x)
3  # Output: 6
```

Computing the Division of two numbers with notation / is performed, as follows:

```
1  x=1/2
2  print(x)
3  # Output: 0.5
```

Computing the Division of two numbers with notation // is performed, where if the result is decimal, their floor is taken, as follows:

```
1  x=1//2
2  print(x)
3  # Output: 0
```

Computing the power of two numbers with notation ** is performed, as follows:

```
1  x=2**3
2  print(x)
3  # Output: 8
```

Computing the minus of two numbers with notation − is performed, as follows:

```
1  x=2−3
2  print(x)
3  # Output: −1
```

In the next, the comparison operators on integers will be described. For the following examples, the result is 'True' or 'False'. For the variables of 'a' and 'b', $a < b$ leads to a true state if a is less than b, as follows:

```
1  a=4
2  b=8
3  z=a<b
4  print(z)
5  # Output: True
```

For the variables of 'a' and 'b', $a > b$ leads to a true state if a is grater than b, as follows:

```
1  a=4
2  b=8
3  z=a>b
4  print(z)
5  # Output: False
```

For the variables of 'a' and 'b', $a <= b$ leads to a true state if a is less than or equal to b, as follows:

```
1     a=4
2     b=8
3     z=b<=a
4     print(z)
5     # Output: False
```

For the variables of 'a' and 'b', $a >= b$ leads to a true state if a is greater than or equal to b, as follows:

```
1     a=4
2     b=8
```

2.2 Data Types in Python

```
3    z=b>=a
4    print(z)
5    # Output: True
```

For the variables of 'a' and 'b', $a == b$ leads to a true state if a is equal to b, as follows:

```
1    a=4
2    b=8
3    z=b=a
4    print(z)
5    # Output: False
```

For the variables of 'a' and 'b', $a! = b$ leads to a true state if a is unequal to b, as follows:

```
1    a=4
2    b=8
3    z=b!=a
4    print(z)
5    # Output: True
```

In the next, the logical operators on integers will be described. For the following examples, the result is 'True' or 'False'. For the variables of 'a' and 'b', a and b leads to a true state if a and b true, as follows:

```
1    a=True
2    b=False
3    z= b and a
4    print(z)
5    # Output: False
```

```
1    a=True
2    b=True
3    z= b and a
4    print(z)
5    # Output: True
```

For the variables of 'a' and 'b', a or b leads to a true state if a or b is true, as follows:

```
1    a=True
2    b=False
3    z= b or a
4    print(z)
5    # Output: True
```

```
1    a=False
2    b=False
3    z= b or a
```

```
4  print(z)
5  # Output: False
```

For the variable of 'a', *not*(*a*) reverse the state 'a', as follows:

```
1  a=False
2  z= not(a)
3  print(z)
4  # Output: True
5  a=True
6  print(z)
7  # Output: False
```

In the above code, the value of variable *a* has changed once.

2.2.3 Decimal Type

Decimal numbers are an important data type in Python, with a wide range of applications in everyday life. They are commonly used in scientific calculations, as well as in calculations involving weights, lengths, times, and financial transactions. Decimal numbers are particularly useful when precision is of utmost importance. For instance, the multiplication of two decimal numbers in Python can be performed using the operator ∗, as shown in the following code:

```
1  x=1.2*3.6
2  print(x)
3  # Output: 4.32
4  ###############
5  x=0.2*4
6  print(x)
7  # Output: 0.8
```

> **Note**
> Note: All operators instructed for the integer numbers can be used for the decimal numbers in Python.

2.2.4 String Type

In Python, strings are a fundamental data type used to represent text. They can contain any character that Python supports and can be written using single quotes,

2.2 Data Types in Python

double quotes, or triple quotes. The use of single or double quotes is interchangeable, with the only difference being that in a single-quoted string, double quotes can be included and vice versa. Triple quotes, on the other hand, enable the creation of strings spanning multiple lines. The following example shows how the strings in Python are used.

```
1   x='Learning'
2   y=''learning''
3   z='''''
4   Learning
5   ''''''
6   W='''
7   Learning
8   '''
```

2.2.4.1 Strings are Iterable and Indexable

In Python, an iterable object is any object that can be looped over using a loop construct such as the for loop. In addition to strings, other iterable objects in Python include lists, sets, tuples, dictionaries, and more.

> **Note**
> Note: Unlike strings, lists, sets, and other iterable objects, integer and decimal types are not iterable in Python. If an attempt is made to loop over an integer or decimal in Python, a TypeError will be raised by the Python interpreter.

String objects can be indexed, which means that individual characters within a string can be accessed and manipulated using their position in the string. Indexing is a common method for addressing elements within iterable objects in Python, and is typically done using a numerical or string-based index. It is important to note that in Python, the index of any object within an iterable is zero-based, meaning that the first element has an index of 0, the second element has an index of 1, and so on. The process of indexing in Python is demonstrated in Fig. 2.4. This figure shows the string 'university' and its corresponding index positions highlighted in red and orange. The red numbers represent direct indexing, where the position of each character is indexed from left to right, starting at 0. The orange numbers represent reverse indexing, where the position of each character is indexed from right to left, starting at −1. In Python, for indexing *objectname*[*n*] is used, where *objectname* is a variable that points to an iterable object, and *n* is a positive or negative integer or

Fig. 2.4 Indexing the string object in Python

-10	-9	-8	-7	-6	-5	-4	-3	-2	-1
U	n	i	v	e	r	s	i	t	y
0	1	2	3	4	5	6	7	8	9

a string. For a string object, n is a positive or negative integer. The following example shows how the strings in Python are indexed.

```
1   x='University'
2   c=x[1]
3   print(c)
4   # output: n
5   c=x[-2]
6   print(c)
7   # t
8   '''
```

In addition to indexing notation, Python also provides a powerful method for addressing one or more elements within an iterable object, known as slicing. In Python, the slicing notation is written as $objectname[n1 : n2 : s]$, where objectname is a variable that points to an iterable object, $n1$ and $n2$ are positive or negative integers representing the start and end indices for slicing, and s indicates the interval between the indices of the iterable object. The following example shows how the strings in Python are sliced.

```
1   x='University'
2   c=x[1:3]
3   print(c)
4   # output: ni
5   c=x[2:7]
6   print(c)
7   # output: ivers
8   c=x[0:7]
9   print(c)
10  # output: Univers
11  c=x[:7]
12  print(c)
13  # output: Univers
14  c=x[4:]
15  print(c)
16  # output: ersity
17  '''''The start and end addresses are not specified,
18  so the whole string object is considered.'''''
19  c=x[::]
20  print(c)
```

2.2 Data Types in Python

```
21   # output: University
22   c=x[::2]
23   print(c)
24   # output: Uiest
25   # Reversing a string
26   c=x[::-1]
27   print(c)
28   # ytisrevinU
29   '''
```

In the above examples, it can be seen that if the start address is not specified, slicing from the zero indexes is started, and if the end address is not specified, slicing from the last index is done. Also, if the start and end addresses are not specified, the whole string is considered. In the above examples, with $x[::2]$ from the zero index until the last index slicing is considered such that just the elements that are located in the even indexes are selected. Additionally, to reverse a string in Python, the slicing notation $x[::-1]$ can be used. This notation extracts the entire string starting from the last index (-1) and moving towards the first index (0) with a step size of 1, effectively reversing the order of the original string.

> **Note**
> Note: The indexing and slicing of all iterable objects in Python is the same.

In the next, the common operations in the strings can be applied with some examples being described. In all examples for the strings, $stringobject$ is a variable that points to a string object. For the variables of $s1$ and $s2$, notation + concat $s1$ and $s2$, as follows:

```
1    s1='learning'
2    s2='data'
3    s3=s1+s2
4    print(s3)
5    # Output: learningdata
6    # defining a string that contains just a space
7    s0='         '
8    s3=s1+s0+s2
9    print(s3)
10   # Output: learning data
```

The operator ∗ can be used to repeat a given string object multiple times. The syntax for repeating a string is *stringobject* ∗ *n*, where stringobject is a variable that points to a string object, and n is a positive integer indicating the number of times to repeat the string. The following example illustrates how operator ∗ is used for the strings.

```
1  s1='learning'
2  s2=s1*3
3  # Output: learninglearninglearning
```

With \n, a new line in the string is generated, as follows:

```
1  s1='#lea234rni' 'ng\nscience'
2  print(s1)
3  # Output: #lea234rning
4  science
```

In the above example, it can be seen that anything that Python encoding supports can be written, and a new line is generated. With \t, a new tab in the string is generated, as follows:

```
1  s1='#lea234rni' 'ng\tscience'
2  print(s1)
3  # Output: #lea234rni' 'ng    science
```

2.2.4.2 String Built-in Functions

> **Note**
> Note: A function is a block of code that performs a specific task and can be called and executed multiple times throughout a program. Functions are a fundamental concept in programming, allowing for the creation of modular, reusable code that can be easily maintained and updated.

> **Note**
> Note: Built-in functions in Python are pre-defined functions that are part of the Python language and can be executed just by referencing their name. These functions are part of the Python Standard Library and provide a wide range of functionality for common programming tasks, such as working with strings, lists, and other data types.

2.2 Data Types in Python

> **Note**
> Note: An argument is any data that is passed to a function when it is called and executed. This data can take many forms, such as integers, strings, lists, tuples, dictionaries, or other objects. A parameter, on the other hand, is a variable defined in the function that is used to receive and process the arguments passed to the function. When calling a function in Python, one or more arguments can be passed to the function as needed, and these arguments are then received by the parameters of function and processed accordingly.

The string object has some built-in functions the most common ones with some examples being described.

2.2.4.3 Count

The *stringobject.count*(g, s, e) (three parameters), counts the number of appearances of g, where *stringobject* is a variable that points to a string object, g is a string object, s determines from what indexing counting is started and e determines until what index counting is continued; s and e are optional, means they can be unused. The following example shows how *count* function is used.

```
1   s='#lea234rni''ng\nscience'
2   c=s.count('i')
3   print(c)
4   # Output: 2
5   c=s.count('i',10)
6   print(c)
7   # Output: 1
8   c=s.count('i',20)
9   print(c)
10  # Output: 0
11  c=s.count('i',2022)
12  print(c)
13  # Output: 0
```

In the above example, it can be seen that if s and e are not specified from the zero index until the last index, counting is performed. The built-in function *count* returns zero if g does not appear in *stringobject* or s and e is the out of range of *stringobject* like *s.count*('i',2022) because the length of s is 21.

> **Note**
> Note: If the start and end indexes are not specified, all indexes of an object are considered.

2.2.4.4 Startswith

The function of *startswith* takes three parameters g, s and e, if *stringobject* starts with g, returns True and else returns False, and the format to use it, is *stringobject.startswith*(g, s, e). The following example shows how *startswith* function is used.

```
1  s='learnscience'
2  c=s.startswith('l',)
3  print(c)
4  # Output: True
5  c=s.startswith('w',)
6  print(c)
7  # Output: False
```

2.2.4.5 Endswith

Function of *endswith* takes three parameters g, s and e, if *stringobject* ends with g, returns True and else returns False, and the format to use it, is *stringobject.endswith* (g, s, e). The following example shows how *startswith* function is used.

```
1  s='learnscience'
2  c=s.endswith('e',)
3  print(c)
4  # Output: True
5  c=s.endswith('w',)
6  print(c)
7  # Output: False
```

2.2.4.6 Find

Function of *find* takes three parameters g, s and e, if *stringobject* find g, returns its index in *stringobject* and else returns -1, and the format to use it, is *stringobject.find*(g, s, e). The following example shows how *find* function is used.

```
1  s='learnscience'
2  c=s.find('t')
3  print(c)
4  # Output: −1
5  c=s.find('i')
6  print(c)
7  # Output: 7
```

2.2 Data Types in Python

```
8   c=s.find('rn')
9   print(c)
10  # Output: 3
```

2.2.4.7 Index

Function of *index* takes three parameters g, s and e, if *stringobject* find g, returns its index in *stringobject* and else raises a ValueError, and the format to use it, is *stringobject.index*(g, s, e). The following example shows how *index* function is used.

```
1   s='learnscience'
2   c=s.index('i')
3   print(c)
4   # Output: 7
5   s='learnscience'
6   c=s.index('m')
7   print(c)
8   # Output: ValueError: substring not found
```

> **Note**
> Note: The only difference between built-in functions of find and index is that if the string is not found find returns −1 while index raises an error.

2.2.4.8 Encoding

Function of *encoding* takes two optional parameters *encoding*, and *errors*, and converts *stringobject* to an encoded sequence. Specifying the type of encoding is with *encoding* parameter, and *errors* determines if a character can not be encoded, what should be replaced instead of the character. By default, the encoding format is $UTF-8$, but it can be changed to other formats. To use *encode*, *stringobject.encode*() is used. The following example shows how *encode* function is used.

```
1   s='LearnScience'
2   c=s.encode()
3   print(c)
4   # b indicates it is a byte string
5   #Output: b'LearnScience'
6   # if a character can not be encoded, it will be ignored.
```

```
 7  c=s.encode(encoding=''ascii'',errors=''ignore'')
 8  print(c)
 9  #Output: b'LearnScience'
10  '''if a character can not be encoded,
11   it will raise a ValueError.
12  '''
13  c=s.encode(encoding=''ascii'',errors=''strict'')
14  print(c)
15  #Output: b'LearnScience'
```

2.2.4.9 Replace

This function replaces a given string with another string given. The *replace* takes two string parameters of *nw* and *pr*, and optional parameter *n*, and *stringobject.replcae(pr, nw, n)*, says how many times *nw* is replaced by *pr* that by default all string that equal to *pr* are replaced. The following example shows how *replace* function is used.

```
1  s='LearnScience'
2  c=s.replace('n','*')
3  print(c)
4  #Output: Lear*Scie*ce
5  s='LearnScience'
6  c=s.replace('n','*',1)
7  print(c)
8  #Output: Lear*Science
```

> **Note**
> Note: The replace function doesn't change the original string.

2.2.4.10 Strip

With *strip*, any specified object is removed at the beginning or the end of string. To use it, *stringobject.strip(chars)* is used where *chars* is an optional parameter that can be any string, by default, chars are space. The following example shows how *islower* function is used.

```
1  s='''as as as    Univer sitY   as s as as'''
2  r=s.strip('as ')
3  print(r)
```

2.2 Data Types in Python

```
4   #Output: as as    Univer sitY    as s as
5   s='''       Univer sitY    '''
6   r=s.strip('as')
7   print(r)
8   #Output:Univer sitY
9   s='''       Univer sitY'''
10  r=s.strip('as')
11  print(r)
12  #Output:Univer sitY
```

2.2.4.11 Join

With *join*, an arbitrary string object to each element of an iterable object (like string and list) is added. To use it, *stringobject.join*(). The following example shows how *join* function is used.

```
1   s1='learning'
2   s2='**'
3   r=s2.join(s1)
4   print(r)
5   #Output: l**e**a**r**n**i**n**g
6   s1='learning'
7   r=''' '''.join(s1)
8   print(r)
9   #Output:l e a r n i n g
10  s1='learning'
11  r='''$b'''.join(s1)
12  print(r)
13  #Output:l$be$ba$br$bn$bi$bn$bg
```

2.2.4.12 Split

With *split*, the string with a separator is splitted into a list. The *stringobject.split-*(*sep, m*) is used to split, where *sep* and *m* are optional, and *sep* is a string that the *stringobject* based on *sep* splitting is done, and *m* determines how many times the split must be done. The following example shows how the *split* function is used. By default, *sep* is a space and *m* is the maximum number of splitting.

```
1   s1='''learning computer science'''
2   r=s1.split()
3   print(r)
4   #Output:['learning', 'computer', 'science']
5   s1='''learning computer science'''
```

```
6   r=s1.split('n')
7   print(r)
8   #Output:['lear', 'i', 'g computer scie', 'ce']
9   s1='''learning computer science'''
10  r=s1.split('n',1)
11  print(r)
12  #Output:['lear', 'ing computer science']
```

In the next, some functions for the string objects are described so that no arguments are sent to them.

2.2.4.13 Capitalize

It *capitalize* the first character (element) of a given string. The following example shows how *capitalize* function is used.

```
1   s='bulGarian'
2   r=s.capitalize()
3   print(r)
4   #Output: BulGarian
5   s='Global'
6   r=s.capitalize()
7   print(r)
8   #Output: Global
```

2.2.4.14 Swapcase

With *swapcase*, if the first character of a given string is a lower case, capitalize it, and if the first character of a given string is a upper case, lower it. The following example shows how *swapcase* function is used.

```
1   s='bulGarian'
2   r=s.swapcase()
3   print(r)
4   #Output: BulGarian
5   s='Global'
6   r=s.swapcase()
7   print(r)
8   #Output: global
```

2.2.4.15 Isalnum

With *isalnum*, if the characters of a given string are alphabet letters, numbers or both, True is returned, else False is returned. The following example shows how *isalnum* function is used.

```
1   s='bulGarian'
2   r=s.isalnum()
3   print(r)
4   #Output: True
5   s='#Global'
6   r=s.isalnum()
7   print(r)
8   #Output: False
9   s='#Global2023'
10  r=s.isalnum()
11  print(r)
12  #Output: True
```

2.2.4.16 Isnumeric

With *isnumeric*, if the characters of a given string are numeric, True is returned, else False is returned. The following example shows how *isnumeric* function is used.

```
1   s=''
2   r=s.isnumeric()
3   print(r)
4   s='202220232024'
5   r=s.isnumeric()
6   print(r)
7   #Output: True
8   s='bulGarian'
9   r=s.isnumeric()
10  print(r)
11  #Output: False
12  s='#Global'
13  r=s.isnumeric()
14  print(r)
15  #Output: False
16  s='#Global2023'
17  r=s.isnumeric()
18  print(r)
19  #Output: False
```

2.2.4.17 Isalpha

With *isalpha*, if the characters of a given string are alphabet letters, True is returned, else False is returned. The following example shows how *isalpha* function is used.

```
1   s='bulGarian'
2   r=s.isalpha()
3   print(r)
4   #Output: True
5   s='#Global'
6   r=s.isalpha()
7   print(r)
8   #Output: False
9   s='#Global2023'
10  r=s.isalpha()
11  print(r)
12  #Output: False
```

2.2.4.18 Islower

With *islower*, if the characters of a given string are in a lower case, True is returned, else False is returned. The following example shows how *islower* function is used.

```
1   s='bulgarian'
2   r=s.islower()
3   print(r)
4   #Output: True
5   s='#global'
6   r=s.islower()
7   print(r)
8   #Output: False
9   s='#Global2023'
10  r=s.islower()
11  print(r)
12  #Output: False
```

2.2.4.19 Isupper

With *isupper*, if the characters of a given string are in an upper case, True is returned, else False is returned. The following example shows how *isupper* function is used.

```
1   s='bulGarian'
2   r=s.isupper()
```

2.2 Data Types in Python

```
3   print(r)
4   #Output: False
5   s='#global'
6   r=s.isupper()
7   print(r)
8   #Output: False
9   s='GLOBAL'
10  r=s.isupper()
11  print(r)
12  #Output: True
```

2.2.4.20 Isspace

With *isspace*, if the characters of a given string are spaces, True is returned, else False is returned. The following example shows how *isspace* function is used.

```
1   s='   '
2   r=s.isspace()
3   print(r)
4   #Output: True
5   s='#global'
6   r=s.isspace()
7   print(r)
8   #Output: False
9   s='world'
10  r=s.isspace()
11  print(r)
12  #Output: False
```

2.2.4.21 Upper

With *upper*, the characters of a given string are changed to upper case. The following example shows how *upper* function is used.

```
1   s='''world'''
2   r=s.upper()
3   print(r)
4   #Output:WORLD
5   s='UniversitY'
6   r=s.upper()
7   print(r)
8   #Output: UNIVERSITY
```

2.2.4.22 Lower

With *lower*, the characters of a given string are changed to lower case. The following example shows how *lower* function is used.

```
1   s='''worLD'''
2   r=s.lower()
3   print(r)
4   #Output:world
5   s='UniversitY'
6   r=s.lower()
7   print(r)
8   #Output: university
9   s='PYTHON'
10  r=s.lower()
11  print(r)
12  #Output: python
```

2.2.4.23 Type Conversion

The process of converting an object from one data type to another is known as type conversion. Python provides built-in functions for converting between different data types such that the type of an integer is represented by *int*, the decimal is represented by *float*, and the string is represented by *str*. Also, the type name of any object in Python with *type(objectname)* is taken. The following examples indicate how to *type* function is used.

```
1   s='''worLD'''
2   r=type(s)
3   print(r)
4   #Output: <class 'str'>
5   s='123'
6   r=type(s)
7   print(r)
8   #Output:<class 'str'>
9   s=4
10  r=type(s)
11  print(r)
12  #Output: <class 'int'>
13  s=8.
14  r=type(s)
15  print(r)
16  #Output: <class 'float'>
17  s=2/4
```

2.2 Data Types in Python

```
18   r=type(s)
19   print(r)
20   #Output: <class 'float'>
```

To convert a value to another type, *int(value)*, *float(value)*, or *str(value)* is used, where *value* is a specified object. Take a look at the following examples to see how the conversion is made.

```
1    c0='4.2'
2    print(type(c0))
3    # Output: <class 'str'>
4    # Converting str to float type
5    c1=float(c0)
6    print(type(c1))
7    # Output: <class 'float'>
8    c0=4
9    print(type(c0))
10   # Output: <class 'int'>
11   # Converting int to float type
12   c1=float(c0)
13   print(type(c1))
14   # Output: <class 'float'>
15   c0=4.2
16   print(type(c0))
17   # Output: <class 'float'>
18   # Converting float to int type
19   c1=int(c0)
20   print(type(c1))
21   <class 'int'>
22   c0='Country'
23   '''
24   An error raises to convert
25   string literals to int or float
26   '''
27   c1=int(c0)
28   print(type(c1))
29   '''
30   Output: ValueError: invalid literal
31   for int() with base 10: 'Country'
32   '''
33   c1=float(c0)
34   print(type(c1))
35   '''
36   Output: ValueError: could not
37   convert string to float: 'Country'
38   '''
```

```
39  c0=80
40  c1=81.9
41  c2=str(c0)
42  c3=str(c1)
43  print(type(c2))
44  print(type(c3))
45  # Output: <class 'str'>
46  # Output: <class 'str'>
```

2.3 Data Structures

Data is one of the most important entities in computer science. Data structures are used to organize, store, and retrieve data efficiently. The choice of data structure determines the format and organization of the stored data, and influences how the data is stored in memory. In the following sections, we will describe some common data structures in Python, including lists, arrays, matrices, sets, and dictionaries, along with examples of how they can be used.

2.4 List

A linked list is a data structure where each element is linked to the next element, or to the next and previous elements, and the elements in memory are not necessarily stored consecutively. In Python, the list data structure is widely used and can be used to implement other data structures such as matrices and arrays. Lists in Python are used to store a sequence of elements, which can be of any data type supported by Python. Lists have the following important features in Python:

1. Storing duplicate elements
   There is no limit to store duplicate elements.
2. Storing heterogeneous elements Elements such as strings, integers, and strings can be stored in a list.
3. Being sortable
   A list can be sorted. For example, sort it from lower to upper, or alphabetically.
4. Being changeable
   Lists are mutable data structures, which means that their content can be modified after they are created. Once data is stored in a list, it can be changed as many times as needed.
5. Being Iterable
   A list is an iterable object that can be indexed.

In Python, lists are created using square brackets [], with the list elements separated by commas inside the brackets. The following examples depict how the list is used.

2.4 List

Fig. 2.5 Example of list structure in Python

$$\begin{bmatrix} -10 & -9 & -8 & -7 & -6 & -5 & -4 & -3 & -2 & -1 \\ U & 4.2 & 8 & v & e & 33 & 7.9 & i & * & 7 \\ 0 & 1 & 2 & 3 & 4 & 5 & 6 & 7 & 8 & 9 \end{bmatrix}$$

```
1  lst=[2,3.1,'−']
2  print(lst)
3  # Output: [2, 3.1, −]
4  print(type(lst))
5  # Output: <class 'list'>
```

In the above example, *lst* is a variable point to a list object that contains the elements of 2, 3.1, '−'. Figure 2.5 depicts an example of a list structure in Python. Alike the string, a list is an object and iterable, so it can be indexed in a direct way and reverse way. Figure 2.5 depicts red numbers are the indexes in a direct way, and orange numbers in a reverse way, and it can be seen that different data types are stored in a list.

2.4.1 List Built-in Functions

The list object in Python provides a number of built-in functions for working with lists, including functions for inserting, removing, counting, indexing, and more. Here are some examples of how these functions can be used. In all examples, *lst* is a variable that points to a list object, *obj* is any object, and *iterobj* is an iterable object.

2.4.1.1 Append

With *append*, an object is inserted into the end of a list each time, To use it, *lst.append(obj)* is used. The following examples indicate how to *append* function is used.

```
1   lst=[2,3.1,'−']
2   lst.append(77)
3   print(lst)
4   # Output: [2,3.1,'−',77]
5   lst0=[101,102,9]
6   lst.append(lst0)
7   print(lst)
8   # Output: [2, 3.1,'−',77 [101, 102, 9]]
9   lst.append(9.9)
10  print(lst)
```

11 # Output: [2, 3.1,'−',77 [101, 102, 9],9.9]

2.4.1.2 Extend

With *extend*, an iterable object is inserted into the end of a list each time. To use it, *lst.extend(iterobj)* is used. With this definition, int object and float do not work with *extend* function. As a solution, float and int objects must be converted to type string and then they can work the *extend* function. The following examples indicate how to *extend* function is used.

```
1  s=''learning''
2  lst=[]
3  lst.extend(s)
4  # Output: ['l','e','a','r','n','i','n','g']
5  s=''learning''
```

In python, if a string is inserted with extend function, each character is inserted into an index. To insert a string into an index of a list with the extend function, the string must split as follows:

```
1  s=''learning''
2  lst=[]
3  lst.extend(s.split())
4  # Output: [''learning'']
```

Other example are as follows:

```
1  lst0=['theta']
2  lst=[]
3  r='learning'
4  lst.extend(lst0)
5  lst.extend(r)
6  print(lst)
7  # Output: ['theta','l','e','a','r','n','i','n','g']
```

The extend() function in Python is used to add elements from an iterable object to the end of a list. However, if a non-iterable object such as an integer or float is passed to the extend() function, Python will raise a TypeError because the object cannot be iterated over.

```
1  lst=[]
2  lst.extend(9)
3  # Output: TypeError: 'int' object is not iterable
4  lst=[]
5  lst.extend(9.87)
6  # Output: TypeError: 'float' object is not iterable
```

2.4 List

In the above examples, Python raises the errors, because int and float are not iterable. Take a look at the following other examples.

```
1  lst=[]
2  lst.extend([9])
3  # Output: [9]
4  lst.append([9])
5  # Output: [[9]]
```

2.4.1.3 Insert

If an object must be inserted into a specific position, the *insert* function must be used. To use it, $lst.insert(ind, obj)$ is used, where ind is an integer that specifies the index that the object must be inserted. The following examples indicate how to *insert* function is used.

```
1   lst=[1,8,9,'8',7,6]
2   lst.insert(-2,'objj')
3   print(lst)
4   # Output: [1, 8, 9, '8', 'objj', 7, 6]
5   lst=[1,8,9,'8',7,6]
6   lst.insert(1,'data')
7   print(lst)
8   # Output: [1, 'data', 8, 9, '8', 'objj', 7, 6]
9   lst.insert(4,7)
10  print(lst)
11  # Output: [1, 'data', 8, 9, 7, '8', 'objj', 7, 6]
```

As another example, consider Code 2.1. When inserting an object in position or index zero, the new object is inserted in the index of zero and the old object is inserted in the index or position one.

Code 2.1 Python code to insert the objects into a list in a reverse order

```
1  current_sublist=[]
2  current_sublist.insert(0,5)
3  current_sublist.insert(0,7)
4  current_sublist.insert(0,220)
5  '''
6  The output is:
7
8  [220, 7, 5]
9  '''
```

If an object is to be inserted at the same index, the previous object should be shifted one position after the current one.

> **Note**
> Note: append, extend, and insert are used to put data into a list, where the append and extend functions put the data to the end of a list each time, and append works with objects, while extend works with iterable objects. The insert function is used to put an object into a specified index.

2.4.1.4 Count

This function is used to count the number of appearances of a given object, if is found the object, the number of its appearance is returned, else zero is returned, to use *lst.count(obj)* is used. The following examples indicate how *count* function is used.

```
1  lst=[1, 'data', 8, 9, 7, '8', 'objj', 7, 6]
2  c=lst.count(7)
3  print(c)
4  # Output: 2
5  c=lst.count('x')
6  print(c)
7  #Output:0
```

2.4.1.5 Index

Function of *index* takes three parameters obj, s and e, if lst find $gobj$, returns its index in lst and else raises a ValueError, The format to use it, is $lst.index(obj, s, e)$, where s is the start address, and e is the end address to find the index. The following example shows how *index* function is used.

```
1  lst=[1, 'data', 8, 9, 7, 8, 7, 6]
2  c=lst.index(8)
3  print(c)
4  # Output: 2
5  c=lst.index(8,3,6)
6  print(c)
7  # Output: 5
```

2.4.1.6 Clear

To clear all content of the list object, *clear* function is used. No argument is sent to *clear* function. The following example indicates how *clear* function can be used.

2.4 List

```
1  lst=[1, 'data', 8, 9, 7, '8', 7, 6]
2  lst.clear()
3  print(lst)
4  # Output: []
```

There is another way to clear a list object.

```
1  lst=[1, 'data', 8, 9, 7, '8', 7, 6]
2  lst=[]
3  print(lst)
4  # Output: []
```

2.4.1.7 Remove

To remove a specific object, *remove* function is used. To use it, *lst.remove(obj)* is used. If the object is not in *lst*, an error is raised. The following examples indicate how *remove* function is used.

```
1  lst=[1, 'data', 8, 9, 7, '8', 7, 6]
2  lst.remove(8)
3  print(lst)
4  # Output: [1, 'data', 8, 9, 7, '8', 7]
5  lst.remove(76)
6  print(lst)
7  # Output: ValueError: list.remove(x): x not in list
```

2.4.1.8 Pop

To remove a specific object, *pop* function is used. To use it, *lst.pop(ind)* is used, where *ind* is an optional parameter that specifies the index object is to be removed. The default value for *pop* is equal to the index of the last item, and if the entered index is not in the range index of *lst*, an error is raised. The following examples indicate how *pop* function is used.

```
1  lst=[1, 'data', 8, 9, 7, '8', 7, 6]
2  lst.pop(8)
3  # Output: IndexError: pop index out of range
4  lst=[1, 'data', 8, 9, 7, '8', 7, 6]
5  lst.pop(3)
6  print(lst)
7  # Output: [1, 'data', 8, 7, '8', 7, 6]
```

> **Note**
> Note: remove and pop are used to delete objects into a list, where remove deletes an object by its object and pop deletes an object by its index. The clear() function is a built-in method of the list object that is used to remove all elements from a list, leaving it empty.

2.4.1.9 Reverse

To reverse the objects of a list, *reverse* function is used. No argument is sent to *reverse* function. The following example indicates how *reverse* function can be used.

```
1  lst=[1, 'data', 8, 9, 7, '8', 7, 6]
2  lst.reverse()
3  # Output: [1, 'data', 8, 9, 7, '8', 7, 6]
```

2.4.1.10 Copy

Consider the following example.

```
1  lst1=[1, 'learn', 8, 9, 7, '8', 7, 6]
2  lst2=lst1
3  lst1.remove('learn')
4  print(lst1)
5  # Output: [1, 8, 9, 7, '8', 7, 6]
6  print(lst2)
7  # Output: [1, 8, 9, 7, '8', 7, 6]
```

In the above example, it can be seen that every change to *lst*1, also will be applied to *lst*2. In fact, with an assignment, a distinct copy is not to be created. To this end, *copy* function is used that has no parameter. The following examples indicate how *copy* function is used.

```
1  lst1=[1, 'learn', 8, 9, 7, '8', 7, 6]
2  lst2=lst1.copy()
3  lst1.remove('learn')
4  print(lst1)
5  # Output: [1, 8, 9, 7, '8', 7, 6]
6  print(lst2)
7  # Output: [1, 'learn', 8, 9, 7, '8', 7, 6]
```

2.4 List   35

2.4.1.11 Nested List

A nested list is a list of lists; or a list/lists are into another list. The following example creates a nested list.

```
1  nlst=[
2  [5.69,8,4.6],
3  ['x','a',0.88],
4  [66, 7.7, '8']]
```

All built-in functions can be applied to the nested lists. Take a look at the following example.

```
1  nlst=[
2  [5.69,8,4.6],
3  ['x','a',0.88],
4  [66, 7.7, '8']]
5  # ind=nlst.index([66, 7.7, '8'])
6  # Output: 2
```

In a nested list, all lists that are in other lists, have their indexes. The more nested, the more indices must be used to access the elements of a nested list. Take a look at the following examples.

```
1   nlst=[
2   [5,8,4],
3   ['x','a',0.88,'r'],
4   [66, 7.7,[10,9,8],4] ]
5   # To access the fist list and its second element
6   val=nlst[1]
7   print(val)
8   # Output: ['x','a',0.88,'r']
9   val=nlst[0][1]
10  print(val)
11  # Output: 8
12  val=nlst[2][3]
13  print(val)
14  # Output: [10, 9, 8]
15  val=nlst[2][2][1]
16  print(val)
17  # Output: 9
```

In the above examples, it can be seen that to access a nested list more than an index must be used. With details, to access element 9 located in [66, 7.7, [10, 9, 8], 4], in the above nested list, [2][2][1] is used, where [2] points to [66, 7.7, [10, 9, 8], 4], [2][2] points to [10, 9, 8], and [2][2][1] points to 9.

2.4.2 List Constructor

Function of *list*() as input takes an optional iterable object, and converts it to a list object, to use it, *list*(*iterobj*) is used. In a simple statement, each element of *iterobj* is placed in an index of the list. Take a look at the following examples. All built-in functions for the list that is created with [] can be applied to the list constructor.

```
1  lst=list('learning')
2  print(lst)
3  # Output: ['l', 'e', 'a', 'r', 'n', 'i', 'n', 'g']
4  lst=list('1234')
5  print(lst)
6  # Output: ['1', '2', '3', '4']
```

2.5 Array

An array consists of data of the same type, whose elements can be accessed using its indices. The elements in an array are stored sequentially in memory as opposed to a list, which means that accessing the elements in the array is faster. In Python, there is no special command to use an array, but a list is used. In other words, as the data in the array must be of the same type, then only the data of the same type is used in the list. In this regard, all built-in functions for the list can be applied to the arrays. Take a look at the following examples.

```
1  # Creating an array of integers with the size of 6
2  arr=[1,4,5,7,9]
3  # Accessing to an element
4  print(arr=[0])
5  # Output: 1
6  # Creating an array of strings with the size of 6
7  arr=['v','c','g','n']
8  d=arr[0]+arr[2]
9  print(d)
10 # Output: vg
```

2.5.1 Sort

Sort is a built-in function of the list, but sort can be applied to data of the same type, so it applies to an array. With $sort(reverse, key)$ function, the elements in ascending or descending order are sorted. In the sort function, $reverse$ and key are optional parameters such that $reverse$ is a binary value, and if it is $False$, sorts in ascending order, else ($True$) sorts in descending order, by default $reverse$ is equal to $False$. The parameter of key determines the sorting criteria. The following examples indicate how $sort$ function is used.

```
1   arr=[3,86,5,7,2,9]
2   # Sorting in ascending order
3   arr.sort()
4   print(arr)
5   # Output: [2, 3, 5, 7, 9, 86]
6   arr.sort(reverse=True)
7   print(arr)
8   # Output: [86, 9, 7, 5, 3, 2]
9   arr=arr.sort(reverse=True)
10  print(arr)
11  # Output: None
```

If the sorting is assigned to a variable, $None$ is returned, where $None$ means no value and has the type of $NoneType$.

2.6 Matrix

A matrix is a two-dimensional data structure in which data is organized into rows and columns. Each element in the matrix is identified by its row and column index. In Python, matrices are commonly implemented using nested lists, where each inner list represents a row of the matrix. To be a valid matrix, each row of the nested list must contain the same number of elements, and the elements should be of the same data type. In this regard, all built-in functions can be applied to the matrices. To address to elements of a matrix, $mat[x][y]$ is used, where mat is a matrix, $[x]$ is used to access the row of a matrix, and $[y]$ is used to access the columns of a matrix. The following examples indicate how the matrix can be used.

```
1   '''''' Creating a matrix 3 * 4 of integers,
2   it has 3 rows, and four columns.
3   The first row: [5,8,4,6],
4   for the first row, column 0, 1,2 and 3 are
5   5,8,4,6 respectively.
6   The second row: [88,4,32,41], for the second row,
7   column 0, 1,2 and 3 are
```

```
 8   88,4,32,41, respectively.
 9   The third row: [66, 3,19,4], for the third row,
10   column 0, 1,2 and 3 are
11   66, 3,19,4, respectively.
12   '''''
13   mat=[
14   [5,8,4,6],
15   [88,4,32,41],
16   [66, 3,19,4] ]
```

```
1   #Creating a matrix 2 * 3 of strings
2   mat=[
3   ['d','f','g'],
4   ['t','g','m'] ]
5   #Creating a matrix 2 * 3 of float type
6   mat=[
7   [5.8,4.6,1.0],
8   [8.8, 3,29.25] ]
```

The following examples are performed for a matrix.

```
 1   mat=[
 2   [5.8,4.6,2.9],
 3   [8.8, 3,29.25] ]
 4   c=mat[1]
 5   print(c)
 6   # Output: [8.8, 3,29.25]
 7   c=mat[1][2]+mat[1][0]
 8   print(c)
 9   # Output: 38.05
10   # All built-in functions can be applied to the matrices too.
11   c=mat[0].count(mat[0][1])
12   print(c)
13   # Output: 1
```

2.6.1 Changing List Elements

In a list, the items can be changed.

```
1   arr=[3,4,9,6,0,330]
2   arr[0]=81
3   arr[4]=23
4   print(arr)
5   # Output: [81, 4, 9, 6, 23, 330]
```

2.7 Set

A set object is a built-in data structure in Python that represents a collection of unique elements. Similar to a list or tuple, a set can contain elements of different data types, and the elements are not stored in any particular order. Some key features of sets in Python include:

1. Non-indexable
   In contrast to a list object, the set object can not be addressed by an index.
2. Unchangeable
   Elements of a set can be added or removed, but not changed by another value.
3. A set cannot have duplicate elements. Sets are used for math, from a practical point of view, sets are used to perform mathematical operations, and remove duplicate elements. To create a set in Python, an open curly bracket '{' and a closed curly bracket '}' are used with initializing value to it, or a *set*() is used.

The following examples indicate how a set object is used. In all examples, *se* is a set object, and *iterobj* is an iterable object.

```
1  se={3,4,7,6,0,9,3,7}
2  print(se)
3  # Output: {0, 3, 4, 6, 7, 9}
4  se={7,6,'learn','a','b','a'}
5  print(se)
6  # Output: {'learn', 6, 7, 'b', 'a'}
```

Notice that with {}, a dictionary is created not a set.

2.7.1 Set Built-in Functions

Set object in Python has some built-in functions that are to be described in the next.

2.7.1.1 Add

With *add* function, an element to a set is added, To use it, *se.add(elem)* is used, where *elem* can not be a list object or set object. The following examples indicate how *add* function in a set is used.

```
1  se={3,4,7,6,0,9,3,7}
2  se.add(21)
3  # Output: {0, 3, 4, 21, 6, 7, 9}
4  se.add(43.4)
5  print(se)
6  # Output: {0, 3, 4, 21, 6, 7, 9, 43.4}
```

```
 7  se.add('eta')
 8  print(se)
 9  # Output: {0, 3, 4, 6, 7, 9, 'eta', 21, 43.4}
10  sl={79,'you'}
11  se.add(sl)
12  # Output: TypeError: unhashable type: 'set'
13  lst=['nba',40,'79']
14  se.add(lst)
15  # Output: TypeError: unhashable type: 'list'
```

2.7.1.2 Update

To include a list or a set *se.update(iterobj)* is used. The following examples indicate how *update* function in a set is used.

```
 1  se={0, 3, 4, 21, 6, 7, 9}
 2  sl={79,'you'}
 3  se.update(sl)
 4  print(se)
 5  # Output: {0, 3, 4, 6, 7, 9, 79, 21, 'you'}
 6  lst=['nba',40,'79']
 7  se.update(lst)
 8  print(se)
 9  '''
10  Output: {0, 3, 4, 6, 7, 'nba', 9, 40, '79', 79, 21, 'you'}
11  '''
```

2.7.1.3 Clear

To clear a set *clear*, *se.clear()* is used. No argument is sent to *clear* function. The following example indicates how *clear* function in a set can be used.

```
1  se={0, 3, 4, 21, 6, 7, 9}
2  se.clear()
3  print(se)
4  # Output: set()
```

2.7.1.4 Union

With *union*, the union of a set with an iterable object is taken, To use it, *se.union (iterobj)* is used. The following examples indicate how *union* function in a set is used.

2.7 Set

```
1   lst=[600,700]
2   stri='abcd'
3   se0={0, 3, 4, 21, 6, 7, 9}
4   se1={89,890}
5   # union of se0 with se1
6   se=se0.union(se1)
7   # Output: {0, 3, 4, 6, 7, 9, 21, 89, 890}
8   # union of se0 with se1
9   se=se1.union(se0)
10  print(se)
11  # Output: {0, 3, 4, 6, 7, 9, 21, 89, 890}
12  # union of a set with a list object
13  se=se0.union(lst)
14  print(se)
15  # Output: {0, 3, 4, 6, 7, 9, 21, 600, 700}
16  # union of a set with a string object
17  se=se0.union(stri)
18  print(se)
19  # Output: {0, 'a', 3, 4, 6, 7, 9, 'b', 21, 'd', 'c'}
```

2.7.1.5 Intersection

With *intersection*, the intersection of a set with an iterable object is taken, To use it, *se.intersection(iterobj)* is used. The following examples indicate how *intersection* function in a set is used.

```
1   lst=[600,700]
2   stri='abcd'
3   se0={0, 3, 4, 21, 6, 7, 9, 600, 'c'}
4   se1={89,890}
5   # intersection of se0 with se1
6   se=se0.intersection(se1)
7   print(se)
8   # Output: set()
9   # intersection of se0 with se1
10  se1={89,890,21}
11  se=se1.intersection(se0)
12  print(se)
13  # Output: {21}
14  # intersection of a set with a list object
15  se=se0.intersection(lst)
16  print(se)
17  # Output: {600}
18  # intersection of a set with a string object
19  se=se0.intersection(stri)
20  print(se)
21  # Output: {'c'}
```

2.7.1.6 Intersection_update

With *intersection_update*, the intersection of a set with k iterable objects is taken, and unlike intersection, a new set is not created, but the changes are applied to the same set. The following examples indicate how *intersection_update* function in a set is used.

```
1   lst=[600,700]
2   stri='abcd'
3   se0={0, 3, 4, 21, 6, 7, 9, 600, 'c'}
4   se1={89,890,4,21}
5   se2={16,166,21}
6   # intersection of se0 with se1 and se2
7   se0.intersection_update(se1,se2)
8   print(se0)
9   # Output: {21}
10  se0.intersection_update(se1,se2,lst)
11  print(se0)
12  # Output: set()
```

2.7 Set

2.7.1.7 Discard

With *discard*, an element of a set is discarded and if the element does not exist, no error is raised, To use it, *discard.intersection(elem)* is used. The following examples indicate how *discard* function in a set is used.

```
1  se={0, 3, 4, 21, 6, 7, 9,600, 'c'}
2  se.discard(4)
3  print(se)
4  # Output: {0, 3, 6, 7, 9, 'c', 21, 600}
5  # 'qr' does not exist in se, but no error is raised.
6  se.discard('qr')
7  print(se)
8  # Output: {0, 3, 6, 7, 9, 'c', 21, 600}
```

2.7.1.8 Remove

With *remove*, an element of a set is removed and if the element does not exist, an error is raised, To use it, *se.remove(elem)* is used. The following examples indicate how *remove* function in a set is used.

```
1  se={0, 3, 4, 21, 6, 7, 9, 600, 'c'}
2  se.discard(4)
3  print(se)
4  # Output: {0, 3, 6, 7, 9, 'c', 21, 600}
5  # 'qr' does not exist in se, so an error is raised.
6  se.discard('qr')
7  print(se)
8  # Output: KeyError: 'qr'
```

2.7.1.9 Pop

With *pop*, an element of a set is randomly removed and the removed item is returned. No argument is sent to *pop* function. The following examples indicate how *pop* function in a set is used.

```
1  se={0, 3, 4, 21, 6, 7, 9, 600, 'c'}
2  d=se.pop()
3  print(d)
4  # Output: 0
```

2.7.1.10 Difference

With $se.difference(iterobj)$, the items that are se and not in $iterobj$ are returned.

```
1  se0 = {0, 3, 4, 21, 6, 7, 9}
2  se1 = {7, 8, 3, 19, 4,'word'}
3  ls1 =[7, 8, 3, 19, 4,'word']
4  se=se0.difference(ls1)
5  print(se)
6  # Output: {0, 21, 6, 9}
7  se=se1.difference(se0)
8  print(se)
9  # Output: {8, 19, 'word'}
```

Notice that the difference also can be applied with − operator, but it just works with the sets, not any iterable objects. Take a look at the following examples.

```
1  se0 = {0, 3, 4, 21, 6, 7, 9}
2  se1 = {7, 8, 3, 19, 4, 'word'}
3  ls1 =[7, 8, 3, 19, 4,'word']
4  se=se0 − se1
5  print(se)
6  # Output: {0, 9, 21, 6}
7  se=se1 − se0
8  print(se)
9  # Output: {8, 19, 'word'}
10 se=se0 − lst
11 print(se)
12 '''
13 Output: TypeError: unsupported operand
14 type(s) for −: 'set' and 'list'
15 '''
```

2.7.1.11 Copy

Consider the following example.

```
1  se1={1, 'learn', 8, 9, 7, '8', 7, 6}
2  se2=se1
3  se1.discard('learn')
4  print(se1)
5  # Output: {1, '8', 6, 7, 8, 9}
6  print(se2)
7  # Output: {1, '8', 6, 7, 8, 9}
```

2.7 Set

In the above example, it can be seen that every change to *se*1, also will be applied to *se*2. In fact, with an assignment, a distinct copy is not to be created (like lists). To this end, *copy* function is used that has no parameter. The following example indicates how *copy* function can be used.

```
1  se1={1, 'learn', 8, 9, 7, '8', 7, 6}
2  se2=se1.copy()
3  se1.discard('learn')
4  print(se1)
5  # Output: {1, '8', 6, 7, 8, 9}
6  print(se2)
7  # Output: {1, '8', 6, 7, 8, 9, 'learn'}
```

2.7.1.12 Issubset

With *se.issubset(iterobj)*, it is checked if all elements of *se* are in *iterobj*, *True* is returned, else *False* is returned.

```
1   lst=[1, 'learn', 8, 9, 7, '8', 7, 6]
2   se2={8,7,9}
3   se3=se2.issubset(lst)
4   print(se3)
5   # Output: True
6   se1={1, 'learn', 8, 9, 7, '8', 7, 6}
7   se3=se2.issubset(se1)
8   print(se3)
9   # Output: True
10  se2={8,7,9,11}
11  se3=se2.issubset(lst)
12  print(se3)
13  # Output: False
```

2.7.2 Set Constructor

Function of *set*() as input takes an optional iterable object, and converts it to a set object, to use it, *set(iterobj)* is used. In a simple statement, each element of *iterobj* is placed in a position of the set. Take a look at the following examples. All built-in functions for the set that is created with {} (with initializing value) can be applied to the set constructor.

```
1  # To remove duplicate elements in a list
2  lst=set(['country', 17, 'x', 17])
```

```
 3  print(lst)
 4  # Output: {'country', 17, 'x'}
 5  # To remove duplicate elements in a string
 6  lst=set('learning')
 7  print(lst)
 8  # Output: {'n', 'i', 'r', 'e', 'l', 'a', 'g'}
 9  lst=set('country')
10  print(lst)
11  # Output: {'n', 'c', 'u', 'r', 't', 'o', 'y'}
12  lst=set(['country',17])
13  print(lst)
14  # Output: {'country', 17}
```

2.8 Dictionary

A dictionary object is another data structure in Python it can store different types of data and when it is used that the pair of data with their attributes are needed, so-called a pair of key-value. For example, it is supposed there is a set of words that is aimed to get their frequency in an arbitrary text. A dictionary is the best data structure to do this, with the keys being the words and the values being the frequencies. The dictionary has the following features.

1. Indexable
   It can access an element with its key, not the index.
2. Changeable
   Elements of a dictionary can be changed by another value.
3. Non-duplicatable
   A dictionary cannot have duplicate elements the same as a set.
4. Nested
   A dictionary can be nested the same as a list.

To create a dictionary in Python, an open curly bracket '{' and a closed curly bracket '}' are used, or the $dict()$ is used. The following examples indicate how the dictionaries in Python are used.

```
1  # Defining an empty dictionary
2  d={}
3  print(type(d))
4  # Output: <class 'dict'>
5  d={'x':4, 1:'9', 'p':'o','x':7, 'x':8}
6  print(d)
7  # Notice that the duplicated keys are removed.
8  # Output: {'x': 8, 1: '9', 'p': 'o'}
```

The indexing in the dictionaries is made as follows.

2.8 Dictionary

```
1  d={'argentina': 1, 'france': 2, 'Croatia': 3, 'morocco': 4}
2  print(d['argentina'])
3  # Output: 1
4  print(d['morocco'])
5  # Output: 4
```

It can be seen that for indexing in a dictionary instead of putting an integer number into the bracket, a key is put. To insert in a dictionary, $dicname[key] = obj$ is used, where $dicname$ is a variable that points to a dictionary object, key is a key belonging to $dicname$, and obj is a Python object. The following examples indicate how inserting in a dictionary is done.

```
1  dic={'argentina': 1, 'france': 2}
2  dic['crotia']=3
3  dic['morocco']=4
4  print(dic)
5  # Output: {'argentina': 1, 'france': 2, 'crotia': 3}
```

Also, the above approach is used to change a key in the dictionary as follows:

```
1  dic={'argentina': 1, 'france': 2, 'crotia': 3}
2  dic['crotia']=2
3  dic['france']=1
4  print(dic)
5  # Output: {'argentina': 1, 'france': 1, 'crotia': 2}
```

2.8.1 Nested Dictionary

A nested dictionary contains other dictionaries inside it. A dictionary can be nested as follows:

```
1  dic={'d': {'argentina': 1, 'Croatia': 3}, 'x':7, 'm2':'no'}
2  print(dic['d'])
3  # Output: {'argentina': 1, 'Croatia': 3}
```

The more nested, the more indices must be used to access the elements of a nested dictionary. For example, to access 'Croatia' and 'argentina', two, and three indexes must be used, respectively.

```
1  di={'d': {'argentina': {'cup':1}, 'Croatia': 3},'x':7}
2  c=di['d']['Croatia']
3  print(c)
4  # Output: 3
5  c=di['d']['argentina']['cup']
6  print(c)
7  # Output:1
```

2.8.2 Dictionary Built-in Functions

Dictionary object in Python has some built-in functions that are to be described with some examples. In all examples, *dic* is a dictionary object, *iterobj* is an iterable object, and *obj* is a Python object.

2.8.2.1 Fromkeys

With *fromkeys*, a specific object is assigned to each element of an iterable object, To use it, *dic.fromkeys(iterobj, obj)* is used. The following examples indicate how *fromkeys* is used.

```
1  d=dict()
2  s='are'
3  c=d.fromkeys(s,12)
4  print(c)
5  # Output: {'a': 12, 'r': 12, 'e': 12}
6  c=d.fromkeys(s,['c'])
7  print(c)
8  # Output: {'a': ['c'], 'r': ['c'], 'e': ['c']}
```

2.8.2.2 Get

Function of *get*, takes two parameters of *key* and *obj* (optional parameter), if the key does not exist, *None* is returned, and if an optional parameter of *obj* is specified and if *key* is not found, *obj* is returned. To use it, *dic.fromkeys(key, obj)* is used. The following examples indicate how *get* is used.

```
1   '''brazil does not exist in the dic dictionary,
2   so {89:'p'} is returned.'''
3   dic=d={'argentina': 1, 'france': 2}
4   c=d.get('brazil',{89:'p'})
5   print(c)
6   # Output: {89: 'p'}
7   '''france exists in the dic dictionary,
8   so its value is returned.'''
9   dic=d={'argentina': 1, 'france': 2}
10  c=d.get('france',8)
11  print(c)
12  # Output: 2
13  dic=d={'argentina': 1, 'france': 2}
14  c=d.get('france')
15  print(c)
16  # Output: 2
```

2.8 Dictionary

2.8.2.3 Update

Function of *update* takes an iterable object that is a pair of keys and values to add to the existing dictionary. Still, the updates are applied to the current dictionary, and a new one is not created. The following example indicates how *update* can be used.

```
1  dic={'argentina': 1, 'france': 2}
2  dic.update(c=[98])
3  print(dic)
4  # Output: {'argentina': 1, 'france': 2, 'c': [98]}
```

2.8.2.4 Clear

To clear a dictionary *dic.clear*() is used. No argument is sent to *clear* function. The following example indicates how *clear* can be used.

```
1  dic=d={'argentina': 1, 'france': 2}
2  d.clear()
3  print(d)
4  # Output: {}
5  print(type(d))
6  # Output: <class 'dict'>
```

2.8.2.5 Items

To obtain all keys and values in a dictionary, *item* function is used, and to use it, *dic.items*() is used. No argument is sent to *items* function. By default, the obtained objects by *items* can not be visible, so to make them visible, they can be converted to a string, and list, set, or dictionary constructor. No argument is sent to *item* function. The following examples indicate how *items* is used.

```
1   dic={'argentina': 1, 'france': 2}
2   d=dict(dic.items())
3   print(d)
4   # Output: {'argentina': 1, 'france': 2}
5   dic=d={'argentina': 1, 'france': 2}
6   d=list(dic.items())
7   print(d)
8   # Output: [('argentina', 1), ('france', 2)]
9   d=set(dic.items())
10  print(d)
11  # Output: {('france', 2), ('argentina', 1)}
12  d=str(dic.items())
13  print(d)
14  # Output: dict_items([('argentina', 1), ('france', 2)])
```

2.8.2.6 Values

To obtain all values in a dictionary, *values* function is used, to use it, $dic.values()$ is used. No argument is sent to *keys* function. By default, the values by *values* can not be visible, so to make them visible, they can be converted to a string, list, or set. No argument is sent to *values* function. The following example indicates how *values* can be used.

```
1  dic={'argentina': 1, 'france': 2}
2  d=list(dic.values())
3  print(d)
4  # Output: [1, 2]
```

2.8.2.7 Keys

With *keys* function, the keys of a dictionary are returned, and to use it, $dic.keys()$ is used. No argument is sent to *keys* function. As the obtained keys by *keys* can not be visible by default, so to make them visible, they can be converted to a string, list, or set. No argument is sent to *keys* function. The following example indicates how *keys* can be used.

```
1  dic={'argentina': 1, 'france': 2}
2  d=list(dic.keys())
3  print(d)
4  # Output: ['argentina', 'france']
```

2.8.2.8 Pop

Function of *pop* takes *key* parameter and optional parameter *obj* to return *key*, if the key does not exist, an error is raised, and if an optional parameter of *obj* is specified and if *key* is not found, *obj* is returned. To use it, $dic.pop(key, obj)$ is used. The following example indicates how *pop* can be used.

```
1  dic={'argentina': 1, 'france': 2}
2  d=dic.pop('iee',8)
3  print(d)
4  # Output: 8
5  d=dic.pop('france')
6  print(d,dic)
7  # # Output: 2 {'argentina': 1}
```

2.8.2.9 Popitem

With *popitem*, the last item in a dictionary is to be removed, and its key and value are to be returned. If the dictionary is empty, an error is raised. To use it, *dic.popitem*() is used. The following example indicates how *popitem* can be used. No argument is sent to *popitem* function.

```
1  dic={'argentina': 1, 'france': 2,'Crotia':3}
2  d=dic.popitem()
3  print(d)
4  # Output: ('Crotia', 3)
5  print(dic)
6  # Dictionary after removing the last item
7  # Output: {'argentina': 1, 'france': 2}
```

2.8.3 Dictionary Constructor

To create a dictionary, *dict*() can be used. The dictionary constructor takes an optional iterable object such that the iterable object must be a pair of keys and values. Take a look at the following examples. All built-in functions for the dictionary that is created with {} can be applied to the dictionary constructor.

```
1   # Defining an empety dictionary
2   d=dict({})
3   print(type(d))
4   d=dict(argentina=1,france=2,Croatia=3)
5   # Output: {'argentina': 1, 'france': 2, 'Croatia': 3}
6   d=dict([('argentina', 1), ('france', 2)], Croatia=3)
7   d=dict(argentina=1,france=2,Croatia=3)
8   # Output: {'argentina': 1, 'france': 2, 'Croatia': 3}
9   d=dict([('argentina', 1), ('france', 2)])
10  print(d)
11  # Output: {'argentina': 1, 'france': 2}
```

2.9 Tuple

In this data structure, the different data types can be stored and can be indexed like a list, but they are not mutable. In other words, once a tuple is created, it cannot be changed such as by adding or removing. The dictionary has the following features.

1. Indexable
   It can access an element with its index.
2. Unchangeable
   Elements of a tuple can not be changed or removed, and no element can be inserted
3. Duplicatable
   A tuple can have duplicate elements the same as a list.
4. Nested
   A tuple can be nested the same as a list.

To create a tuple in Python, an open parenthesis (, and a closed parenthesis) are used, or a *tuple*() is used.

> **Note**
> Note: As a tuple is immutable, there is no need to copy it, just an assignment is enough.

The following examples indicate how the dictionaries in Python are used.

```
1  # Defining an empty tuple
2  t=()
3  print(type(t))
4  # Output: <class 'tuple'>
5  t=('x', 4, 1, '9', 'p', 'o', 'x', 9, 7, 'x', 8)
6  print(t)
7  # Notice that the duplicated keys are kept.
8  # Output: ('x', 4, 1, '9', 'p', 'o', 'x', 9, 7, 'x', 8)
```

A tuple can be created without parentheses, only the elements must be separated by commas. Take a look at the following example.

```
1  t='argentina', 1, 'france', 2, 'Croatia', 3, 'morocco', 4
2  print(t)
3  '''
4  Output: ('argentina', 1, 'france', 2,
5  'Croatia', 3, 'morocco', 4)
6  '''
```

The indexing in the tuples is done as follows.

```
1  t=('argentina', 1, 'france', 2, 'Croatia', 3, 'morocco', 4)
2  print(t[0])
3  # Output: argentina
4  print(t[3])
5  # Output: 2
```

2.9 Tuple 53

2.9.1 Nested Tuple

A nested tuple contains other tuples inside it. The more tuple, the more index to access an object.

```
1  # defining a nested tuple
2  t=('argentina', 1, 'france',('gold','silver',(2,4)))
3  # Accessing to elements of a nested tuple
4  ind=t[3][2][1]
5  print(ind)
6  # Output: 4
```

In the above example, index $t[3][2][1]$ is used to access an element in tuple object t, where [3] points to (*'gold'*, *'silver'*, (2, 4)), [2] points to (2, 4), and [1] points to 4.

2.9.2 Tuple Built-in Functions

Tuple object in Python has some built-in functions that are to be described with some examples. In all examples, *tup* is a tuple object, *iterobj* is an iterable object, and *obj* is a Python object.

2.9.2.1 Count

This function is used to count the number of appearances of a given object, if it is found the object, the number of its appearance is returned, else zero is returned, to use *tup.count(obj)* is used. The following examples indicate how *count* function is used.

```
1  t=('argentina', 1, 'france',[99],1)
2  y=t.count(1)
3  # Output: 2
```

2.9.2.2 Index

To use index, $t.up.index(obj)$ is used.

2.9.3 Tuple Constructor

With *tuple*, an iterable object is converted to a tuple that to use it, *tuple(iterobj)* is used. Take a look at the following examples.

```
1   t=tuple('abcs=d')
2   print(t)
3   # Output: ('a', 'b', 'c', 's', '=', 'd')
4   t=tuple('abcs=d')
5   print(t)
6   # Output: ('a', 'b', 'c', 's', '=', 'd')
7   t=tuple([7,6,3,0])
8   print(t)
9   # Output: (7, 6, 3, 0)
10  t=tuple({'x':1, 'y':2})
11  print(t)
12  # Output: ('x', 'y')
13  t=tuple({'x', 'y'})
14  print(t)
15  # Output: ('y', 'x')
```

2.9.4 Range Type

The range type creates a sequence of integers, which is typically used to iterate through for loops. The range type takes integer *start* parameter, and optional integer parameters of *stop* and *step*, where *start* specifies the number that the range starts, *stop* specifies the number that range is ended by it, and *step* indicates in what interval the numbers have appeared. If the optional parameters are not specified, the range is stated from zero, and length range is equal to *start* − 1 (as the indexes are started from zero). By default, the numbers in the range object are not visible to make them visible, they must be converted to a list, tuple, or set. The following examples indicate how the *range* in Python is used.

```
1   x=range(10)
2   print(type(x))
3   # Output: <class 'range'>
```

```
1   '''
2   Converting a range object to a list in an incremental way
3   '''
4   x=list(range(10))
5   print(x)
6   # Output: [0, 1, 2, 3, 4, 5, 6, 7, 8, 9]
7   # Converting a range object to a set
8   x=set(range(10))
9   print(x)
10  # Output: {0, 1, 2, 3, 4, 5, 6, 7, 8, 9}
11  # Converting a range object to a tuple
```

```
12   x=tuple(range(10))
13   print(x)
14   # Output: (0, 1, 2, 3, 4, 5, 6, 7, 8, 9)
```

In the above examples, the *stop* and *step* are not specified, so the range is started from zero to 10 − 1. Take a look at the other examples to include *stop* and *step* in the range object.

```
1    '''
2    Creating a range of numbers in range
3    10 to 16 in an incremental way
4    x=list(range(10,17))
5    print(x)
6    # Output: [10, 11, 12, 13, 14, 15, 16]
7    # A range of numbers in range 10 to 16 by step 2
8    x=list(range(10,17,2))
9    print(x)
10   # Output: [10, 12, 14, 16]
11   # A range of numbers in range 10 to 16 by step 3
12   x=list(range(10,17,3))
13   print(x)
14   # Output: [10, 13, 16]
```

Also, the range function can generate integer numbers in a decremental way, as follows.

```
1    # Creating a range of numbers in range 17 to 11
2    x=list(range(17,10,−1))
3    print(x)
4    # Output: [17, 16, 15, 14, 13, 12, 11]
5    # Creating a range of numbers in range 17 to 11
6    x=list(range(17,10,−2))
7    print(x)
8    # Output: [17, 15, 13, 11]
```

2.10 Statements

The concept of line in Python can be interpreted in two ways: Physical lines refer to the visible lines of code that are written in a Python script or program. These lines are typically separated by line breaks and can be seen when viewing the code in a text editor or integrated development environment (IDE). Logical lines, on the other hand, are the interpreter's interpretation of the statement. A logical line may consist of one or more physical lines.

A statement in Python is an instruction that is executed by the interpreter. Simple statements are typically written on a single physical or logical line, but multiple simple statements can be written on a single line if they are separated by semicolons (;). However, it is generally not recommended to write multiple statements on a single line, as it can make the code difficult to read and maintain. Take a look at the following examples.

```
1   '''
2   ''a=10'' is a physical line and logical line
3   that is a simple statement
4   '''
5   a=10
6   '''
7   ''b=13; a=34'' is a physical line and two logical lines
8   that is a simple statement
9   '''
10  b=13; a=34
```

All the statements discussed in previous sections are simple statements, which are executed one after another. However, in Python, the indentation of a line of code is used to define a block of code, rather than using braces or other symbols like in some other programming languages. A block of code in Python is defined by using indentation. All the statements within a block must be indented with the same amount of white spaces relative to the header. This is in contrast to languages like C++, where blocks of code are defined by using braces. Another important concept in Python is a compound statement, which is composed of several simple statements that affect the execution of the program. Compound statements typically start with a header that consists of a keyword followed by an expression, and end with a colon (:). The statements that follow the header are indented with the same amount of white spaces and are considered to be part of the block of code.

Some examples of compound statements in Python include the if statement for conditional branching, the while and for loops for iterating over a sequence of values, and the def keyword for defining functions.

```
1   # Simple statement
2   lst=[1, 'learn', 8, 9, 7, '8', 7, 6]
3   # Simple statement
4   se2={8,7,9}
5   # Simple statement
6   se3=se2.issubset(lst)
7   # Simple statement
8   print(se3)
```

In the above example, it can be seen that all statements are simple and have the same amount of indention. The following example illustrates how compound statements are written.

```
1   '''
2   header is consisted with if keyword and expression 20>18,
```

2.10 Statements

```
3    and is ended by : (colon)
4    '''
5    if 20 > 18 :
6        '''
7        it can be seen that there
8        is some indentation relative
9        to the header
10       '''
11       x=0
12       x=x+1
13       print(''greater'')
14       '''
15       All statements after the
16        header are in the same
17        amount of indentation
18       '''
```

```
1    x=0
2    n=20
3    '''
4    header is consisted
5    with while keyword
6    and expression x <=n,
7    and is ended by : (colon)
8    '''
9    while x <=n:
10       x=x+1
11       print(n)
12       '''
13       All statements after the
14       header are in the same
15       amount of indentation
16       '''
```

```
 1    '''
 2    header is consisted with
 3    for keyword and an iterable object,
 4    and is ended by : (colon)
 5    '''
 6    lst=[2,5,7,9,3]
 7    for i in lst:
 8        x=0
 9        print(i)
10        '''
11        All statements after the header
12        are in the same amount of indentation
13        '''
```

2.11 Conditions

In a program, there is a need for some statements to be executed under certain conditions usually. In programming languages, the if structure is used to make some conditions. The following description indicates how the 'if' structure is used. First, the first 'if' header is checked, and if the condition was True, the statements are executed by a Python interpreter, if the first 'if' is not satisfied, the second 'elif' (means else if) is considered, and if the first 'elif' is not satisfied, the second 'elif' is considered. This process is repeated until all 'elif' ones are checked, and if no 'elif' was not satisfied, the statements in the 'else' header are executed. In Python, the 'if' structure is a compound statement such that 'if', 'elif', 'else' are headers, and the statements below them are some indentations relative to the header, and they must have the same amount of indentation. Notice that the 'elif', and 'else' are optional.

```
 1    '''
 2    if condition :
 3        statement 1
 4        statement 2
 5        statement n
 6    elif condition:
 7        statement 1
 8        statement 2
 9        statement n
10    .
11    .
12    .
13    elif condition:
14        statement 1
15        statement 2
```

2.11 Conditions

```
16       statement n
17   else:
18       statement 1
19       statement 2
20       statement n
21   '''
```

In the next, some examples of the 'if' structure in Python is given.

```
1   '''
2   Determining the maximum number
3    from three given numbers
4   '''
5   x=431
6   y=238
7   z=437
8   maximum = 0
9   if x > y and x > z: # Header
10      maximum = x
11  elif y > x and y > z: # Header
12      maximum =y
13  elif z > x and z > y: # Header
14      maximum = z
15  print('''The maximum number is ', maximum''')
16  # Output: The maximum number is 437
```

In the below code, a simple type checker with an 'if' structure is provided.

```
1   # Type checkr
2   x='learn'
3   if type(x)==int: # Header
4       print('''the type is integer''')
5   elif type(x)==float: # Header
6       print('''the type is float''')
7   else: # Header
8       print('''''''the type is string''''''')
9   # Output: the type is string
```

Also, an 'if' structure can be nested. Take a look at the following examples:

```
1   x='learn'
2   if type(x)==int:
3       print('''the type is integer''')
4   elif type(x)==float:
5       print('''the type is float''')
6   else:
7       if x=='learning':
8        print('''your strings ends with ing''')
```

```
 9      elif x=='learn':
10          print('''It is matched''')
11          # Output: It is matched
```

2.12 Loops

Loops are used when a block of code needs to be executed repeatedly for a specific purpose. There are two main types of loops in Python: the while loop and the for loop. In Python, 'for' is a compound statement, so all statements after the header have the same amount of indentation relative to the header of 'for'.

2.12.1 For

The following notes depict the structure of 'for' loop in Python.

```
 1
 2
 3    ''' The 'in' is a keyword that is
 4     used to check the existence
 5     of a object in to a sequence
 6     ( range, list, string, etc)'''
 7    '''
 8    # sequence can be range, list, set, dict, string, etc.
 9    for variable in sequence: # header
10        statement 1
11        statement 2
12        .
13        .
14        .
15        statement n
16    '''
```

The following example takes a list of numbers and determines whether they are odd or even.

```
1   x=[1,5,8,9,109]
2   for i in x:
3     if i %2==0:
4       print('''the number'''+ str(i)+''' is even''')
5     else:
6       print('''the number'''+ str(i)+''' is odd''')
7   '''
```

2.12 Loops

```
8    the number 1 is odd
9    the number 5 is odd
10   the number 8 is even
11   the number 9 is odd
12   the number 109 is odd
13   '''
```

In a 'for' loop, the number of variables in the variable part can be more than one, in case the objects of the iterable object in an index be the same as the number of variables. Take a look at the following examples to see how it works.

```
1    for i,j in ([(1,9),(8,9),(7,6),(7,6)]):
2        print(i,j)
3        '''
4        Output:
5        1 9
6        8 9
7        7 6
8        7 6
9        '''
10   for i,j,k in ([(1,9,'c'),(8,9,'s'),(7,6,'d')]):
11       print(i,j,k)
12       Output:
13       1 9 c
14       8 9 s
15       7 6 d
16       '''
```

There is another example of using for loop and dictionary.

```
1    dic={'cs':10,'ds':13,'phd':4,'ms':10}
2    for i,j in dic.items():
3        print(i,j)
4        '''
5        Output:
6        cs 10
7        ds 13
8        phd 4
9        ms 10
10       '''
```

Also, the 'for' loops can be nested, the next example shows how with nested loops a matrix is created.

```
1    # Creating a matrix 5 * 6 that is filled with zero
2    mat = []
3    for i in range(5):
4        row = []
```

```
5      for j in range(6):
6          row.append(0)
7      mat.append(row)
8  print(mat)
9  '''
10 Output:
11 [[0, 0, 0, 0, 0, 0],
12  [0, 0, 0, 0, 0, 0],
13  [0, 0, 0, 0, 0, 0],
14  [0, 0, 0, 0, 0, 0],
15  [0, 0, 0, 0, 0, 0]]
16 '''
```

2.12.2 Comprehensions

Comprehensions in Python are faster in terms of time and shorter code to create a sequence of data. The following example is a way to create a sequence of data in a list.

```
1  lst=[]
2  for x in range(6):
3      lst.append(x)
4  print(lst)
5  # Output: [0, 1, 2, 3, 4, 5]
```

The following code is an example of creating a sequence of data with a list comprehension. It can be seen that the code is shorter than the above code, and actually it is faster.

```
1  # List comprehension
2  lst=[x for x in range(6)]
3  print(lst)
4  # Output: [0, 1, 2, 3, 4, 5]
```

Also, many different operations with a list comprehension can be done.

```
1  # List comprehensions
2  lst=[x+1 for x in range(2)]
3  print(lst)
4  # Output: [1, 2]
5  lst=[x-1 for x in range(4)]
6  print(lst)
7  # Output: [-1, 0, 1, 2]
8  lst=[x**2 for x in range(6)]
9  print(lst)
```

2.12 Loops

```
10   # Output: [0, 1, 4, 9, 16, 25]
11   lst=[x*3 for x in range(3)]
12   print(lst)
13   # Output: [0, 3, 6]
```

The comprehension concept can be applied to a dictionary and set too. Take a look at the following examples.

```
1    # Set comprehension
2    u={x**2 for x in range(7)}
3    print(u)
4    # Output: {0, 1, 4, 36, 9, 16, 25}
5    # Dictionary comprehension
6    u={x:x-1 for x in range(7)}
7    print(u)
8    # Output: {0: -1, 1: 0, 2: 1, 3: 2, 4: 3, 5: 4, 6: 5}
9    # Dictionary comprehension
10   u={'''value '''+str(x):x*1.5 for x in range(7)}
11   print(u)
12   # Output: {'value 0': 0.0, 'value 1': 1.5, 'value 2': 3.0}
```

The 'while' loop when is used when the number of iterations is not known before iterating. In fact, the 'while' loop each time is iterated if its condition is satisfied. In Python, 'while' is a compound statement, so all statements after the header have the same amount of indentation relative to the header of 'for'. The following note depicts the structure of 'while' loop in Python.

```
1    '''
2    while expression: #header
3         statement 1
4         statement 2
5         .
6         .
7         .
8         statement n
9    '''
```

Take a look at the following example.

```
1    # Copying n to x
2    x=0
3    n=20
4    while x<n:
5         x=x+1
6    print(x)
7    # Output: 20
```

The another example is to sort the given numbers in an ascending order.

```
1   unsorted_list=[8,7,5,9]
2   for indexi in range(1, len(unsorted_list)):
3       currentkey=unsorted_list[indexi]
4       prevind=indexi-1
5       while prevind>-1 and currentkey<unsorted_list[prevind]:
6           unsorted_list[prevind+1]=unsorted_list[prevind]
7           prevind=prevind-1
8       unsorted_list[prevind+1]= currentkey
9   print(unsorted_list)
```

2.13 Functions

A function is a block of code that performs a specific task and usually returns a result. There are two main types of functions: built-in functions and user-defined functions. Built-in functions are pre-defined functions that can be executed simply by referencing them, whereas user-defined functions are defined and executed by the user. There are many built-in functions available, and investigating all of them is beyond the scope of this study. However, we will describe some of them with examples below. In the below examples, $iterobj$ is an iterable object. Taking the maximum of the given numbers is with $max(iterobj)$.

```
1   t=[4,7,9,3]
2   # Taking the maximum number from a list
3   m=max(t)
4   print(m)
5   # Output: 9
6   # Taking the maximum number from a set
7   t={4,7,9,3}
8   m=max(t)
9   print(m)
10  # Output: 9
11  # Taking the maximum number from a tuple
12  t=(4,7,9,3)
13  m=max(t)
14  print(m)
15  # Output: 9
```

Taking the summation of the given numbers is with $sum(iterobj)$.

```
1   t=[4,7,9,3]
2   # Taking the maximum number from a list
3   s=sum(t)
4   print(s)
5   # Output: 23
```

2.13 Functions

```
 6   # Taking the maximum number from a set
 7   t={4,7,9,3}
 8   s=sum(t)
 9   print(s)
10   # Output: 23
11   # Taking the maximum number from a tuple
12   t=(4,7,9,3)
13   s=sum(t)
14   print(s)
15   # Output: 23
```

With $abs(n)$ (n is a number) a value is taken as a parameter and returns its absolute value.

```
1   t=abs(-81)
2   print(t)
3   # Output: 81
4   t=abs(-4.1)
5   print(t)
6   # Output: 4.1
7   t=abs(6)
8   print(t)
9   # Output: 6
```

The enumerate function is used to iterate over an iterable object while keeping track of both the elements and their indices. To use it, $enumerate(iterobj, s)$ is used, where $iterobj$ is an iterable object, and start specifies the starting index (default is zero). By default, the objects in the enumerate object are not visible. However, you can convert it to a tuple, list, set, or dictionary to make the objects visible.

```
 1   l1=['x',1,9,'cs']
 2   t1=('x',1,9,'cs')
 3   ''' Converting a list to an enumerable object,
 4   and to be visible, it is converted to a list.
 5   '''
 6   enum=list(enumerate(l1))
 7   print(enum)
 8   # Output: [(0, 'x'), (1, 1), (2, 9), (3, 'cs')]
 9   # The indexes are started from 14
10   enum=set(enumerate(l1,14))
11   print(enum)
12   # Output: {(17, 'cs'), (16, 9), (15, 1), (14, 'x')}
13   # The indexes are started from 4
14   enum=tuple(enumerate(l1,4))
15   print(enum)
16   # Output: ((4, 'x'), (5, 1), (6, 9), (7, 'cs'))
17   # The indexes are started from 4
```

```
18  enum=dict(enumerate(l1,4))
19  print(enum)
20  # Output: {4: 'x', 5: 1, 6: 9, 7: 'cs'}
21  # Converting a tuple to an enumerable object
22  enum=list(enumerate(t1))
23  print(enum)
24  # Output: [(0, 'x'), (1, 1), (2, 9), (3, 'cs')]
```

With *len(seq)* function, as input a sequence is taken, and its length is returned. The following example indicates how *len* function can be used.

```
1   r=[1,4,9,8,5]
2   t=len(r)
3   print(t)
4   # Output: 5
5   s={1,4,7,2}
6   print(len(s))
7   # Output: 4
8   st='ComputerScience'
9   print(len(st))
10  # Output: 15
11  print(len(2))
12  # Output: TypeError: object of type 'int' has no len()
13  print(len(8.5))
14  # Output: TypeError: object of type 'float' has no len()
```

In the above examples, it can be seen that the parameter or parameters as input are taken, and a result is returned. The enumerate function in the for loops is widely used. The following example indicates, how it is used.

```
1   # Output: considering indexes and objects together
2   s='''cs ds'''
3   for j in enumerate(s):
4       print(j)
5       ''' Output:
6       (0, 'c')
7       (1, 's')
8       (2, ' ')
9       (3, 'd')
10      (4, 's')
11      '''
12  # Indexes are started from 2
13  s='cs'
14  for j in enumerate(s,2):
15      print(j)
16      '''
17      # Output:
```

2.13 Functions

```
18      (2, 'c')
19      (3, 's')
20      '''
```

To obtain the object and the index separately in the for loop, two variables are used as follows.

```
1   s='cs'
2   for i,j in enumerate(s,2):
3       print(i,j)
4   '''
5   # Output:
6   2 c
7   3 s
8   '''
```

The above built-in functions were written by others, but Python users themselves can create functions called user-define functions. The following notes depict how the user-defined function is defined. In Python, user-defined functions are the compound statement, which means that after defining the header, all statements after the header have the same amount of indentations.

```
1   ''''''
2   def is a keyword,
3   FunctionName is the name that the user chooses,
4   parameter is to import the data from the arguments,
5   return is the result that
6   inside the function is computed.
7   ''''''
8   '''
9   # Header
10  def FuncName (parameter1, parameter2,... ,parametern):
11      statemnt 1
12      statement 2
13      .
14      .
15      .
16      statement n
17      return result
18  '''
```

In Python, if no argument is sent to the function, its default parameters are considered, otherwise, the arguments sent are used. Take a look at the following example.

```
1   # Parameters are operator', operrand1, operand2
2   def SimpleCalc(operator='+',operrand1=20, operand2=16):
3       if operator=='+':
4           res=operrand1+operand2
```

```
 5      elif operator=='*':
 6          res=operrand1*operand2
 7      elif operator=='-':
 8          res=operrand1-operand2
 9      elif operator=='/':
10          res=operrand1/operand2
11      return res
12  # No argument is sent.
13  t=SimpleCalc()
14  print(t)
15  # Output: 36
```

In the above example, although no parameter is sent to *SimpleCalc* function, it has parameters that are initialized with the values. Hence, no error is raised. If the arguments are passed to *SimpleCalc*, the default parameters are ignored, as follows.

```
 1  # Parameters are operator', operrand1, operand2
 2  def SimpleCalc(operator='+', operrand1=20, operand2=16):
 3      if operator=='+':
 4          res=operrand1+operand2
 5      elif operator=='*':
 6          res=operrand1*operand2
 7      elif operator=='-':
 8          res=operrand1-operand2
 9      elif operator=='/':
10          res=operrand1/operand2
11      return res
12  # arguments of '*',12,2 are sent to SimpleCalculator
13  t=SimpleCalc('*',12,2)
14  print(t)
15  # Output: 24
```

2.13.1 Variable Scopes

The variable that points to the objects of a string, set, list, object, etc, its scope can be local or global. The Local variables are only used within functions, while global variables can be used throughout the program. For example, the following code will raise an error, because *r* is a local variable and just can be used in the function.

```
1  def add(a,v):
2      c=a+v
3      r=4
4      r=r*v
5      return c
```

2.13 Functions

```
6  d=add(11,9)
7  print(r)
8  # Output: NameError: name 'r' is not defined
```

With *global* keyword, the local variable inside a function can be used as a global.

```
1  def add(a,v):
2      c=a+v
3      global r
4      r=4
5      r=r*v
6      return c
7  d=add(11,9)
8  print(r)
9  # Output: 36
```

The next example indicates how the numbers are sorted using a user-defined function.

```
1  def insertion_sort(unsorted_list):
2    for indexi in range(1, len(unsorted_list)):
3      currentkey=unsorted_list[indexi]
4      PreInd=indexi-1
5      while PreInd>-1 and currentkey <unsorted_list[PreInd]:
6        unsorted_list[PreInd+1]=unsorted_list[PreInd]
7        PreInd=PreInd-1
8      unsorted_list[PreInd+1]= currentkey
9    return unsorted_list
10 t=insertion_sort([4,7,9,0,3,31])
11 print(t)
12 # Output: [0, 3, 4, 7, 9, 31]
```

2.13.2 Lambda Function

The lambda function is a simple way to define a function in Python. It is often called an anonymous function because it has no name. Lambda functions are useful when a small, simple function is needed and there is no need to define a separate function using the def keyword. A lambda function can take any number of arguments, but it can only evaluate one expression. The syntax for a lambda function is lambda arguments: expression. The arguments are separated by commas, and the expression is evaluated and returned when the lambda function is called. Unlike regular functions, there is no need to use the return keyword in a lambda function because it automatically returns the result of the expression. The following note depicts the structure of 'lambda' in Python.

```
1  '''
```

```
2  # arg means argument
3  lambda: arg1, arg2, argn: expression
4  '''
```

Take a look at the following examples.

```
1  t=lambda a,b: a+b
2  r=t(5,9)
3  print(r)
4  # 14
```

```
1  t=lambda a,b: a*b
2  r=t(5,9)
3  print(r)
4  # 45
```

2.13.3 Handling the Control Flow

In some cases, unwanted situations such as unwanted objects or errors need to be handled. In such cases, different ways of handling the situation can be considered, such as exiting from the loop or ignoring the error. The loop can be exited using the break keyword when a certain condition is met. This is useful when it is necessary to stop iterating through the loop after a certain point. The continue keyword can be used to skip over a certain iteration of a loop and move on to the next iteration. This is useful when it is necessary to skip over certain items in a loop. The pass keyword is used to do nothing and is typically used as a placeholder when it is necessary to leave a function or code block empty. The try and except statements are used to catch and handle exceptions in Python. When an error occurs in a try block, the code in the corresponding except block is executed to handle the error and prevent the program from crashing.

> **Note**
> Note: The *break* and *continue* just work if they are used in the loops.

To see how they work, take a look at the following examples.

2.13.3.1 Continue

In the below example, with *continue* keyword, if a number is even it is ignored, and the loop until the end is continued.

2.13 Functions

```
1  lst=[5,6,2,3,0,13,4]
2  lst0=[]
3  for i in lst:
4      if i%2==0:
5          continue
6      lst0.append(i)
7  print(lst0)
8  # Output: [5, 3, 13]
```

2.13.3.2 Break

In the below example, with *break* keyword, if just an even number is found, the loop is terminated, which 6 is found, and the loop is terminated.

```
1  lst=[5,6,2,3,0,13,4]
2  lst0=[]
3  for i in lst:
4      if i%2==0:
5          break
6      lst0.append(i)
7  print(lst0)
8  # Output: [5]
```

With *pass* keyword, nothing is done, as follows.

```
1  lst=[5,6,2,3,0,13,4]
2  lst0=[]
3  for i in lst:
4      if i%2==0:
5          pass
6      lst0.append(i)
7  print(lst0)
8  # Output: [5, 6, 2, 3, 0, 13, 4]
```

2.13.3.3 Try Except

When running a program, errors may occur that prevent the program from running. In Python, errors can be handled with *try* and *except* structures. It is a compound statement, which means all statements after the header have the same amount of indentation. The following note depicts how *try* and *except* work.

```
1  '''
2  try: # Header
3      statement 1
```

```
4      statement 2
5      .
6      .
7      .
8      statement n
9  except ExceptionName 1: # Header
10     statement 1
11     statement 2
12     .
13     .
14     .
15     statement n
16  except ExceptionName 2: # Header
17     statement 1
18     statement 2
19     .
20     .
21     .
22     statement n
23  except ExceptionName 3: # Header
24     statement 1
25     statement 2
26     .
27     .
28     .
29     statement n
30  '''
```

In Python, errors can occur for various reasons, including value, type, index, etc. Hence, there are different exceptions to handle each error. Consider the following example to see how it works.

```
1   '''
2   In a try–except block, each statement can
3   be written within the try block and will
4   be executed unless an exception is raised.
5   '''
6   a=10
7   b=21
8   try:
9       c=a+b+'b'
10      print(c)
11  # If a value error occurs,
12  # an alert will be raised through printing
13  except ValueError:
14      print('''Value Error''')
```

2.14 Modules

```
15  # If an index error occurs,
16  # it does nothing
17  except IndexError:
18      pass
19  # If a Zero Division Error is raised,
20  # an appropriate action will be taken
21  # to handle the erro
22  except ZeroDivisionError:
23      print('''zero error''')
24  # If a Type Error occurs,
25  # it will print a and b
26  except TypeError:
27      print(a,b)
```

Specifying the *ExceptionName* is optional, and if there is no *ExceptionName*, all types of errors with one *except* are handled.

```
1  lst=[1,4,7,9,8]
2  try:
3      c=lst[8]
4      print(c)
5  except:
6      pass
```

In the above example, an index error occurs, but just with the *except* can be handled. In fact, it is used *except* without specifying the exception name when just handling the error is essential.

2.14 Modules

A module is a collection of code that contains a set of predefined functions that can be used by importing. To import a module, the import keyword is used, followed by the name of the module, *import libname* that *import* is a keyword, and *libname* is a library name. For example, the math module can be imported using import math. Once a module is imported, the functions and variables defined in the module can be accessed using the dot notation. For example, the sqrt() function from the math module can be accessed using math.sqrt(4).

The following examples indicate how the libraries are imported. The math library is used for importing the math functions.

```
1  import math
2  # Computing cosine
3  t=math.cos(2.8)
4  print(t)
5  # Output: −0.9422223406686581
```

```
6   # Computing logarithm in base 2
7   t0=math.log2(43)
8   print(t0)
9   # Output: 5.426264754702098
10  # Computing logarithm in base 10
11  t1=math.log10(43)
12  print(t1)
13  # Output: 1.6334684555795864
14  # Computing factorial
15  t2=math.factorial(5)
16  print(t2)
17  # Output: 120
18  '''
19  There are other functions in the math library,
20  just some of them are described.
21  '''
```

The random library is used to generate random values for objects.

```
1   import random
2   '''
3   Generating a random number in range (a,b):
4   a and b are integer numbers
5   '''
6   c=random.randint(3,6)
7   print(c)
8   # Output: 4
9   print(c)
10  # Output: 6
11  # shuffle a list
12  lst=[3,9,7]
13  random.shuffle(lst)
14  print(lst)
15  # [7, 9, 3]
16  '''
17  There are other functions in the random library,
18  just some of them are described.
19  '''
```

2.15 Generators

Generators in Python are functions that return an iterator instead of generating a value. They are useful when dealing with a large amount of data, as they allow for loading only the necessary part of the data into memory, instead of loading the

2.16 Recursion

entire dataset. The syntax for generators is similar to that of functions, with the key difference being the use of the "yield" keyword instead of "return". To illustrate this concept, consider the Python code in Code 2.2. This code employs a generator to produce an iterator containing 10 billion numbers. If this were done using a regular function, it would result in a memory error.

Code 2.2 An example of generators in python

```
1  def gen1(n):
2      a=0
3      number=0
4      while a<n:
5          a=a+1
6          yield number
7          number=number+1
8  s=gen1(10000000000)
```

In Code 2.2, as long as $a < n$, $yiled$ generated a number.

2.16 Recursion

In computer science, recursion is a technique wherein a function repeatedly calls itself until a certain condition is met (base case). The significance of recursion lies in its ability to efficiently solve complex problems that can be decomposed into smaller, similar sub-problems. The following code computes the factorial of a number using recursion.

```
1   #calculating the factorial of a number n using recursion
2   def factorial(n):
3
4       # Check if n is equal to 0 (base case)
5       if n==0:
6           # Return 1 as the factorial of 0
7           return 1
8
9
10      #If n is not equal to 0,
11      #calculate the factorial recursively
12
13      else:
14          # Multiply n by the factorial of n−1
15          return n * factorial(n−1)
```

Also, the following code computes the nth Fibonacci number using recursion.

```
'''
Calculating the n-th Fibonacci
number using recursion
'''
def fibonacci(n):

    '''
    Check if n is less than
    or equal to 1 (base case)
    '''
    if n <= 1:
        # Return n as the Fibonacci number
        return n
    '''
    If n is greater than 1, calculate
    the Fibonacci number recursively
    '''
    else:
        # Add the two previous Fibonacci numbers
        return fibonacci(n-1) + fibonacci(n-2)
```

Chapter 3
Math

This chapter pertains to 28 mathematical problems, which are elucidated through a series of examples and subsequently implemented in Python. The enumerated list of these problems is presented below:

1. Computing the execution order in Josephus problem
2. Reaching the point (0,0) in the lattice path
3. Generating a sorted list of integers for the Brussel choice problem:
4. Finding the inverse of collatz conjecture
5. Counting the possible corners
6. Nearest s-gonal for an integer n
7. Finding the fulcrum position of physical weights
8. Computing the total number of blocks required to build a pyramid of spheres
9. Grouping the identical coins based on some conditions
10. Computing the median of the triple medians
11. Smallest seven-zero
12. Converting and Evaluating Mathematical Expressions from Postfix to Infix Notation
13. Reaching the stable state in the Bulgarian solitaire
14. Computing the area of the rectangular towers in the Manhattan Skyline
15. Cutting the rectangular into the squares
16. Eliminating the corners in the two-dimensional grid
17. Computing Last row of the Leibniz Triangle
18. Reaching to a goal based on Collatzy distance
19. Equaling the summation of two squares with n
20. The summation of three numbers with a goal number
21. Finding perfect power
22. Lunar multiplication of integers
23. n-th term of recaman sequence

24. n-th term of Van Eck sequence
25. Generating Non-Consecutive Fibonacci Sums Based on Zeckendorf's Theorem
26. Finding the k-th Fibonacci word
27. Finding the most point line in a two-dimensional grid
28. Checking if a centrifuge is balanced.

3.1 Josephus

Josephus is a theoretical problem in computer science. The problem involves a group of people standing in a circle, awaiting execution. After the first person is executed, a certain number of people are skipped, and another person is executed. This process of execution continues around the circle until only one person remains. The individuals who have been executed are not included in the removal process, and the last person standing does not take their own life. The objective is to develop a function that, given values for n and k, returns the order in which the individuals will be executed. Here, n represents the total number of people, and k represents the number of skips. It is important to note that k can be smaller than, equal to, or even greater than n.

For example, consider Fig. 3.1, where there are five individuals arranged in a circle. If $k = 2$ and $n = 5$, output the desired execution order. Let ex be an array to store the men to be executed. As $k = 2$, after two steps x_2 is found, x_2 is removed and not included in the circle; $ex = [2]$. In each step, k must incremented two by two, so $2 + 2$, $k = 4$, and x_4 is removed and not included in the circle; $ex = [2, 4]$. Again, k is incremented, so x_1 is found, and is removed; $ex = [2, 4, 1]$. Notice that

Fig. 3.1 An example of people that are around a circle

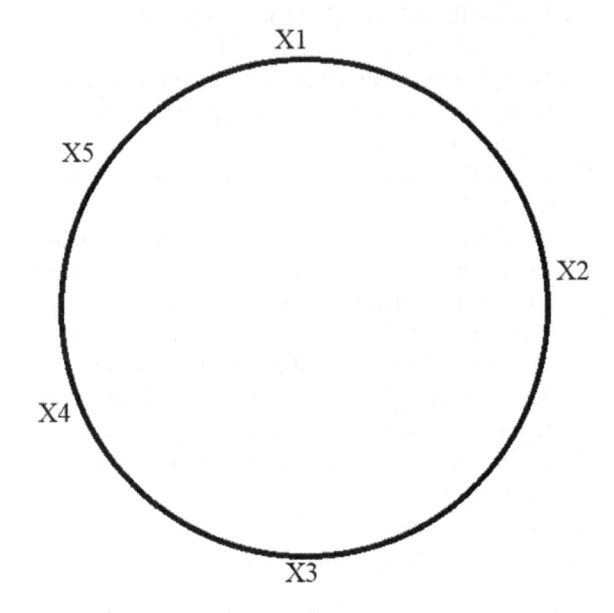

3.1 Josephus

Table 3.1 The expected outputs for certain inputs for Josephus problem

n, k	Expected output
6, 2	[2, 4, 6, 3, 1, 5]
5, 2	[2, 4, 1, 5, 3]
8, 8	[8, 1, 3, 6, 5, 2, 7, 4]
3, 9	[3, 1, 2]
4, 3	[3, 2, 4, 1]

the indexes are reset, as it is a circle, so in the previous step, $k = 1$. Now in the set of men, there are x_3 and x_5, and $k = 1$. K is incremented by two, so x_5 is found, and removed, so $ex = [2, 4, 1, 5]$. In the last, x_3 is executed, and in the execution order is $ex = [2, 4, 1, 5, 3]$. For some inputs, the expected outputs are illustrated in Table 3.1.

Algorithm

It takes two arguments: n, the number of people in the circle, and k, the number of people to skip before eliminating the next person. It creates a list m of the numbers from 1 to n, representing the people in the circle. The algorithm then repeatedly eliminates the k-th person from the circle until only one person remains, and returns a list of the order in which the people were eliminated. It uses the modulo operator to ensure that the index i wraps around to the beginning of the list when it reaches the end. The algorithm steps are outlined in detail as follows:

1. It takes two values n and k.
2. Create a list named m that contains all the numbers from 1 to n.
3. Create an empty list named ans to store the output.
4. Set the variable i to 0.
5. Create a while loop that will continue until the list 'm' becomes empty.
6. Within the while loop, by utilizing Eq. 3.1, update i by adding k to it and subtracting 1. Then take the modulo of the length of m to ensure that i remains within the bounds of the list.
7. Use the pop method on the m list to remove the element at index i and append it to the ans list.
8. Return the ans list as the output of the function.

Code 3.1 depicts the Python code for Josephus's problem.

$$(i + k - 1) \mod m \tag{3.1}$$

Code 3.1 Python code for Josephus problem

```python
def josephus(n, k):
    '''
    Create a list of the numbers from 1 to n,
    representing the people in the circle.
    '''
    m = list(range(1, n + 1))

    '''
    Create an empty list to store the order
    in which the people are eliminated.
    '''
    ans = []

    '''
    Initialize a variable i to keep track of
    the current position in the list.
    '''
    i = 0

    '''
    Loop until there are no more
    people left in the circle.
    '''
    while m:
        # Calculate the index of the person to eliminate.
        i = (i + k - 1) % len(m)

        '''
        Pop the person from the list
        and add them to the elimination order.
        '''
        ans.append(m.pop(i))

    # Return the elimination order.
    return ans
```

Josephus function operates as follows:

It generates a list, m, that includes all the numbers from 1 to n. An empty list, ans, is created to store the executed people. The variable i is instantiated to 0 which represents the starting index. A while loop is entered and continues until all the people have been executed. Within the loop, the function calculates the index of the next number to execute by adding k − 1 to the current index i and taking the result modulo the length of the remaining list m. It then appends the number at the calculated index to the ans list and removes it from the m list using the pop method. Finally, the function returns the ans list containing the order in which the people were executed.

3.2 Reaching a Point in Lattice Path

Let the point (x,y) be in a lattice grid that consisted of the pairs of natural numbers, and one want to reach to point (0,0). At each point, the movement must be either left or down. Write a function that counts the highest number of states that can be passed so that the point does not exceed the coordinates in the taboo list. As input, x and y are the coordinates and *tabu* is the list of forbidden points. For some inputs, the expected outputs are illustrated in Table 3.2.

Algorithm

The algorithm is as follows:

1. It takes row, col, and tabu.
2. Create a 2D array paths of size (row+1) × (col+1)
3. Initialize paths[0][0] to 1
4. For all i from 1 to row, if (i, 0) is not in tabu, set paths[i][0] to paths[i−1][0]
5. For all j from 1 to col, if (0, j) is not in tabu, set paths[0][j] to paths[0][j−1]
6. For all i from 1 to row and for all j from 1 to col, if (i, j) is not in tabu, set paths[i][j] to paths[i−1][j] + paths[i][j−1]
7. Return paths[row][col]

Table 3.2 The expected outputs for certain inputs for Lattice Path

x, y, tabu	Expected output
3, 2, []	10
1, 6, [(7, 1), (4, 4)]	7
8, 8, [(9, 10), (1, 4)]	11220
7, 5, [6,8]	792

Code 3.2 depicts the Python code for the Lattice path problem.

Code 3.2 Python code for lattice path problem

```python
def lattice_paths(row, col, tabu):
    '''Create a 2D array with size (row+1) x (col+1)
    and initialize all values to zero
    '''
    paths = [[0] * (col+1) for _ in range(row+1)]

    '''Set the value of the top-left corner cell to 1,
    since there is only one way to reach it
    '''
    paths[0][0] = 1

    '''Fill in the first column of the table,
    skipping any cells that are in the 'tabu' list '''
    for i in range(1, row+1):
        if (i, 0) not in tabu:
            paths[i][0] = paths[i-1][0]

    '''Fill in the first row of the table,
    skipping any cells that are in the 'tabu' list
    '''
    for j in range(1, col+1):
        if (0, j) not in tabu:
            paths[0][j] = paths[0][j-1]

    # Fill in the rest of the table
    for i in range(1, row+1):
        for j in range(1, col+1):
            # Skip any cells that are in the 'tabu' list
            if (i, j) not in tabu:
                '''The value of each non-border cell
                    is the sum of the value of the cell
                    above it and the cell to the left of it.'''
                paths[i][j] = paths[i-1][j] + paths[i][j-1]

    return paths[row][col]
```

3.3 Brussel Choice Problem

In this challenge, a function must be created for the Brussel choice problem. The function takes positive integers n, mink, and maxk as input parameters and generates a sorted list of all positive integers n for which there exists a subset m that can take

3.3 Brussel Choice Problem

Table 3.3 The expected outputs for certain inputs for Brussel Choice

n, mink, maxk,	Expected output
10, 2, 5	[5, 20]
9, 1, 4	18
47, 1, 1	[27, 87, 414]
100, 84, 99	[]

on values of either 2 times m or the fraction m/2. The resulting list is required to be in ascending order while ensuring that the subset m falls within the range of mink and maxk. For some inputs, the expected outputs are illustrated in Table 3.3.

Algorithm

In general, the algorithm creates a numerical list through specific operations performed on the digits of the input number 'n'. The said operations comprise dividing a subset of digits by 2 or multiplying it by 2 based on whether the subset represents an even or odd integer. Subsequently, the resultant numerical list is sorted in ascending order. The algorithm steps are outlined in detail as follows:

1. The algorithm takes three parameters: 'n', 'mink', and 'maxk'.
2. Initialize an empty list called 'result'.
3. Convert the input integer 'n' to a list of its digits using list comprehension and save it to the variable 'digits'.
4. Iterate over 'k' from 'mink' to 'maxk+1'.
5. For each value of 'k', iterate over 'i' from 0 to 'len(digits)-k+1'.
6. Extract the sub-list of digits from index 'i' to 'i+k-1', convert it to an integer and save it to the variable 'digit'.
7. Check if 'digit' is even. If so, compute half of 'digit' and save it to the variable 'half_digit'.
8. Create a new list called 'new_digits' which consists of the first 'i' elements of 'digits', followed by the digits of 'half_digit', followed by the remaining elements of 'digits'.
9. Convert 'new_digits' to an integer and save it to the variable 'new_num'.
10. Append 'new_num' to the list 'result'.
11. Create a new integer called '*nd*' which is twice 'digit'.
12. Create a new list called 'new_digits' which consists of the first 'i' elements of 'digits', followed by the digits of 'new_digit', followed by the remaining elements of 'digits'.
13. Convert 'new_digits' to an integer and save it to the variable 'new_num'.
14. Append 'new_num' to the list 'result'.
15. Sort the list 'result' using the insertion sort algorithm in an ascending order.
16. Return the sorted list 'result'.

Code 3.3 depicts the Python code for the Brussel choice problem.

Code 3.3 Python code for brussels choice problem
```python
def insertion_sort(arr):
    for i in range(1, len(arr)):
        k = arr[i]
        j = i - 1
        while j >= 0 and k < arr[j]:
            arr[j + 1] = arr[j]
            j -= 1
        arr[j + 1] = k
    return arr

def brussels_choice_problem(n, min_k, max_k):
    res = []
    digits = [int(d) for d in str(n)]
    for k in range(min_k, max_k+1):
        for i in range(len(digits)-k+1):
            d = int(''.join(str(d)
                    for d in digits[i:i+k]))
            if d % 2 == 0:
                hd = d // 2
                new_digits = digits[:i] + \
                    [int(d) for d in str(hd)] + digits[i+k:]
                new_num = int(''.join(str(d)
                                for d in new_digits))
                res.append(new_num)
            nd = d * 2
            new_digits = digits[:i] + \
                [int(d) for d in str(nd)] + digits[i+k:]
            new_num = int(''.join(str(d)
                            for d in new_digits))
            res.append(new_num)
    res = insertion_sort(res)
    return res
```

3.4 Inverse Collatz Conjecture

The Collatz conjecture is a mathematical statement that asserts that any integer smaller than 2^{68}, satisfying the two conditions specified in Eq. 3.2, will eventually reach the number 1 after a finite number of steps. The conjecture deals with sequences and can be expressed as follows: Begin with a natural number n, then generate two

3.4 Inverse Collatz Conjecture

Table 3.4 The expected outputs for certain inputs for inverse collatz conjecture

Shape	Expected output
'dd'	4
'uudduuudd'	None
'ududududdddudddd'	15
'uduuddduddduu'	None

numbers for each number according to Eq. 3.2. Repeat this process for all subsequent numbers until the sequence ends with 1.

Your task in this challenge is to write a function that computes the inverse of the Collatz conjecture. If the computation yields an invalid result (i.e., a non-integer number), your function should return None. The inverse of the Collatz conjecture doubles even numbers and subtracts 1 and divides odd numbers by 3. Formally, the Collatz conjecture is represented in Eq. 3.2, while the inverse of the Collatz conjecture is represented in Eq. 3.3. For certain inputs, expected outputs are provided in Table 3.4. For example, if $shape = ududddd$ is the given string, compute the inverse of the collatz conjecture. In the first step, it visits d, according to Eq. 3.3, multiplication is done, so $x = x \times 2$. In the next step, three consecutive d are visited that x is updated as follows, $x = 2 \rightarrow 2*2 = 4$, $x = 4 \rightarrow 4*2 = 8$, $x = 8 \rightarrow 8*2 = 16$. In the next step, u is visited and according to Eq. 3.3, division is done, so $x = 16 \rightarrow (16-1)/3 = 5$, again, a d is visited and $x = 5 \rightarrow 5*2 = 10$. In the last, one u is visited, so $x = 10 \rightarrow (10-1)/3 = 3$. Therefore, the inverse of collatz conjecture for $shape = ududddd$ is 3. To check if the obtained inverse is correct, the shape of the given string in the inverse of collatz conjecture must be equal to its shape in collatz conjecture. In this regard, according to Eq. 3.2, while x is not reached to 1, the string is traversed to obtain the shape. From the previous step, $x = 3$ and let t be an empty string; hence, $num = x$, $(num \mod 2) = 1$, so $num = (3*num + 1) \rightarrow num = (3*3 + 1)$, and $t = u$. Again, $num = 10$, $10 \mod 2 = 0$, so $10/2 = 5$, and $t = ud$. In the next step, $num = 5$, and is still unequal to 1, so $num/mod2 = 1$, and $num = 5*3 + 1 = 16$, and $t = udu$. The other steps are as follows: $num = 16 \mod 2 = 0 \rightarrow num = num/2 = 8 \rightarrow t = udud$, $num = 8 \mod 2 = 0 \rightarrow num = num/2 = 4 \rightarrow t = ududd$, $num = 4 \mod 2 = 0 \rightarrow num = num/2 = 2 \rightarrow t = udddd$, $num = 2 \mod 2 = 0 \rightarrow num = num/2 = 1 \rightarrow t = ududddd$. Now, num is equal to 1, so the checking process, if the given string in the inverse of collatz conjecture, and collatz conjecture are equal, is finished. To this end, t is equal to $shape$, which means, the obtained inverse of collatz conjecture is correct.

$$x = \begin{cases} x/2 & if\ x\%2 == 0 \\ 3x + 1 & if\ x\%2 == 1 \end{cases} \quad (3.2)$$

$$x = \begin{cases} 2x & if\ x\%2 == 0 \\ (x-1)/3 & if\ x\%2 == 1 \end{cases} \quad (3.3)$$

Algorithm

The algorithm to solve this problem is to transform an input string comprising solely of 'u' and 'd' characters into a numerical value using the inverse of the Collatz conjecture. To begin with, the algorithm initializes x as 1.0 and proceeds to process each character in the input string from right to left. If the current character is 'd', the algorithm doubles x. Conversely, if the character is 'u', it reduces x based on the inverse of the Collatz conjecture. In case the preceding value of x is not an integer, the algorithm returns None. Subsequently, once all characters have been processed, the algorithm verifies that x is neither equal to zero nor empty. Finally, the algorithm computes the expected outcome by applying the Collatz conjecture to the generated number and compares it with the original input string. If both match, the algorithm returns the computed number as an integer. Otherwise, it returns None. The Python code for collatz conjecture is depicted in Code 3.4.

Code 3.4 Python code for inversing collatz conjecture

```
 1  def check_if_integer(number):
 2      '''
 3      Check if a number is an integer by
 4      checking the remainder when divided by 1
 5      '''
 6      if number % 1 == 0:
 7          return True
 8      else:
 9          return False
10
11  def pop_last_item(input_list):
12      # Get and remove the last item from the input list
13      list_length = len(input_list)
14      last_item = input_list[list_length-1]
15      del input_list[list_length-1]
16      return last_item
17
18  def Inverse_collatz_conjecture(shape):
19      '''
20      Convert the input to a list of
21      characters and initialize x to 1.0
22      '''
23      shape_list = list(shape)
24      x = 1.0
25
26      # Iterate over each character in the input list
27      while shape_list:
28          item = pop_last_item(shape_list)
29          if item == 'd':
```

3.4 Inverse Collatz Conjecture

```
30              '''Double the value of x if the current item
31              is 'd'
32              '''
33              x *= 2
34          elif item == 'u':
35              '''Decrease the value of x according to the
36              inverse of Collatz conjecture if the current
37              item is 'u'
38              '''
39              prev = (x - 1) / 3
40              is_integer = check_if_integer(prev)
41
42              '''Check if the previous value is an integer,
43              if so update x
44              '''
45              if is_integer:
46                  x = prev
47              else:
48                  # If the value is not an integer, return None
49                  return None
50
51      # Check for zero or empty input
52      if x == 0 or not shape:
53          return None
54
55      true_answer = ''
56      num = x
57
58      '''
59      Calculate the correct answer
60      using the Collatz conjecture
61      '''
62      while num != 1:
63          if num % 2 == 0:
64              true_answer += 'd'
65              num /= 2
66          elif num % 2 == 1:
67              true_answer += 'u'
68              num = 3 * num + 1
69
70      # Compare the calculated and expected answers
71      if true_answer == shape:
72          return int(x)
73      else:
74          return None
```

3.5 Counting Possible Corners

On a two-dimensional integer grid, a corner is defined as three points of the form (x, y), (x, y + h), and (x + h, y) for some value of h greater than 0. These points form a carpenter's square or a chevron shape that points towards the southwest, with the point (x, y) serving as the tip and (x, y + h) and (x + h, y) defining its axis-aligned wings of equal length. Write a function that as input take a list of points sorted by their x-coordinates, ties broken by y-coordinates, and returns the number of possible corners. For some inputs, the expected outputs are illustrated in Table 3.5.

Algorithm

The algorithm to count the possible corners is to find all 3-point combinations in the list of points that form a carpenter's square or chevron shape as described in the problem statement. The algorithm steps are outlined in detail as follows:

1. initialize a variable named *counter* to zero.
2. Iterate over each point in the list of points using a nested loop.
3. For each point, check if there exists another point with a larger x-coordinate and the same y-coordinate.
4. If such a point exists, check if the third point needed to form a corner (as per the definition provided in the problem) is also present in the list of points.
5. If the third point exists, increment the counter variable by 1.
6. Return the final value of the *counter*.

The Python code for counting possible corners is depicted in Code 3.5.

Code 3.5 Python code for counting corners

```
1  def counting_possible_corners(points):
2      counter = 0
3      for x, y in points:
4          for x2, y2 in points:
5              if x2 > x and y2 == y and \
6                 (x + y2 - y, y + x2 - x) in points:
7                  counter += 1
8      return counter
```

Table 3.5 The expected outputs for certain inputs for counting possible corners

Points	Expected output
[(1, 1), (3, 5), (5, 2)]	4
[(0, 4), (0, 16), (2, 2), (2, 5), (5, 2), (9, 13)]	1
[(1, 3), (1, 7), (5, 3), (5, 5), (7, 3)]	15

3.6 Nearest S-gonal

Table 3.6 The expected outputs for certain inputs for finding nearest S-gonal

n, sides	Expected output
7, 8	8
1, 19	1
15, 18	18
87, 36	105

3.6 Nearest S-gonal

Let $s > 2$ be a positive integer that defines the infinite sequence of s-gonal numbers, where the i-th element is represented by Eq. 3.4. Write a function that as input takes sides s and a positive integer n, and return the Nearest s-gonal integer to n. If there are two s-gonal numbers, the smaller one must be returned. For some inputs, the expected outputs are illustrated in Table 3.6.

$$\frac{((s-2) \times i^2) - ((s-4) \times i)}{2} \tag{3.4}$$

Algorithm

In order to find the polygonal number closest to a given number 'n' with a specified number of sides 'sides'. The algorithm follows a three-step process. Firstly, it computes the polygonal number for a specific index by employing the equation Eq. 3.4. Secondly, it utilizes binary search to identify the middle index that corresponds to the polygonal number closest to 'n'. Lastly, the distances between the polygonal numbers and 'n' are calculated for the three indices that surround the middle index obtained through binary search, and the polygonal number that is nearest to 'n' is returned.

The Python code for finding nearest polygonal number is depicted in Code 3.6.

Code 3.6 Python code for finding nearest s-gonal number

```
1  def nearest_polygonal_number(n, sides):
2      # Calculating polygonal numbers
3      def calculate_polygonal_number(index):
4          return ((sides - 2) *
5              index ** 2 - (sides - 4) * index) // 2
6
7      # Determine upper bound for binary search
8      upper_bound = 1
9      while calculate_polygonal_number(upper_bound) <= n:
10         upper_bound *= 2
11
```

```python
# Perform binary search to find middle index
lower_bound, upper_bound = 0, upper_bound
while lower_bound < upper_bound:
    middle_index = (lower_bound + upper_bound) // 2
    if calculate_polygonal_number(middle_index) < n:
        lower_bound = middle_index + 1
    else:
        upper_bound = middle_index

'''
In Python, inf is a floating-point
    value representing positive
infinity. It is a special
floating-point value that represents
a number greater than any other finite
value, including other floating-point
values and integers.
'''
closest_distance = float('inf')
'''
Calculate distances and find
nearest polygonal number
'''
nearest_polygonal = None
for i in [-1, 0, 1]:
    polygonal = \
    calculate_polygonal_number(lower_bound + i)
    distance = abs(polygonal - n)
    if distance < closest_distance:
        closest_distance = distance
        nearest_polygonal = polygonal

# Return the nearest polygonal number
return nearest_polygonal
```

3.7 Finding Fulcrum Position

Term 'Fulcrum' refers to the point of balance within a list of weights, where the total weight on its left side is equal to the total weight on its right side. This challenge requires identifying a position in a non-empty list of numerical values that can serve as the fulcrum and balance the weight on both sides. According to the principles of physics, a board achieves equilibrium when the forces acting on either side of its

3.7 Finding Fulcrum Position

Table 3.7 The expected outputs for certain inputs for finding fulcrum positions

Weights	Expected output
[6, 6, 9]	−1
[43, 51, 35, 4]	1
[19, 25, 5, 42, 38, 8, 34, 16, 14, 8, 47, 42, 4, 20, 23]	7
[7, 24, 3, 38]	2

support are equal. To meet this objective, you must create a function that takes a list of numbers as input and returns the position of the fulcrum that balances the weights as output. In cases where no such position exists, the function should return −1.

Algorithm

To locate the fulcrum's position in a non-empty list of numerical values, which balances the weights on each of its sides. The algorithm gets a list of numbers as input, iterates through each index in the list, determines the cumulative weight on either side of a point for each index, and outputs the index where the fulcrum maintains balance (i.e., where the sum of weights on the left side matches the sum of weights on the right side). If no such position exists, the algorithm returns −1.

Outlined below are the detailed steps involved in the algorithm:

1. taking a list of numbers known as 'weights'.
2. Use a for loop with the range function to iterate through the input list's indices.
3. For each index 'i' within the loop, compute the sum of weights on the left-hand side of the fulcrum (defined as the elements before index 'i') by iterating through those elements and multiplying them by their distance from the point (i.e., i−j). Also, determine the sum of weights on the right-hand side of the fulcrum (defined as the elements after index 'i') by iterating through those elements and multiplying them by their distance from the point (i.e., j−i).
4. If the left and right weight sums are equal (i.e., the fulcrum is balanced), return the current index 'i' as the position of the fulcrum.
5. If no such position exists, return −1.

By following the above-stated steps, the algorithm can locate the position of the fulcrum that balances the weights in the input list. For some inputs, the expected outputs are illustrated in Table 3.7.

The Python code for finding the fulcrum position is depicted in Code 3.7.

Code 3.7 Python code for finding fulcrum position

```
1  def find_fulcrum_position(weights):
2      for i in range(len(weights)):
3          left_weight_sum = sum(weights[j] * (i-j)
```

```
4        for j in range(i))
5            right_weight_sum = sum(weights[j] * (j-i)
6        for j in range(i+1, len(weights)))
7            if left_weight_sum == right_weight_sum:
8                return i
9    return -1
```

3.8 Counting Sphere Pyramid Blocks

In this challenge, there is a pyramid constructed using spheres. The pyramid has n rows and m columns of spheres on the top level, and each subsequent level has one additional row and column of spheres compared to the previous level. You are requested to write a function that takes in the dimensions n and m as well as the height h, and computes the total number of blocks required to build such a pyramid. For some inputs, the expected outputs are illustrated in Table 3.8.

Algorithm

To calculate the total number of spheres required, the algorithm utilizes a set of formulas that take into account the number of rows and columns in each level of the pyramid, as well as the height of the pyramid. Outlined below are the detailed steps involved in the algorithm:

1. Subtract 1 from the input height 'h'. The bottom-most layer is counted as the first layer and the top-most layer is counted as the $(h + 1)$th layer. However, the formula used in the code assumes that the bottom-most layer is the 0th layer and the top-most layer is the h-th layer. Therefore, to make the height 'h' consistent with the formula used, we need to subtract 1 from the input height 'h' (i.e., 'h +1 − 1' or 'h').
2. Compute the total number of blocks required using the following formula: $a = m * n * (h + 1)$

Table 3.8 The expected outputs for certain inputs for problem of spheres in pyramids

n, m, h	Expected output
4, 2, 2	23
28, 30, 3	2699
3, 3, 3	50
2, 7, 9	654

3.8 Counting Sphere Pyramid Blocks

3. Add the extra blocks required for each level of the pyramid, using the following formula: $a\mathrel{+}= (m+n)*(h*(h+1)//2)$. This calculates the sum of the first h triangular numbers, multiplied by the sum of *m* and *n*.
4. Finally, add in the extra blocks required to create each layer of the pyramid, using the following formula: $a\mathrel{+}= (h*(h+1)*(2*h+1))//6$.
5. Return the final value of *a*. This calculates the sum of the first h squared numbers, divided by 6.

The Python code to compute the total number of spheres required to build a pyramid is depicted Code 3.8.

Code 3.8 Python code for counting the number of spheres in pyramids

```
def Counting_sphere_pyramid_blocks(n, m, h):
    h -= 1
    '''
    Formula to compute the total number
    of blocks required to build
    a pyramid with base dimensions
    n by m and height h.
    '''

    a = m * n * (h + 1)
    a += (m + n) * (h * (h + 1) // 2)
    a += (h * (h + 1) * (2 * h + 1)) // 6
    return a
```

3.9 Grouping Coins

In this challenge, an array of n identical coins is given and needs to be divided into g groups, where g is a positive integer greater than or equal to one. If there are any coins left over after forming complete groups, the remaining coins in the array are set aside and retained. To calculate the number of coins needed to form each group, we use the formula Eq. 3.5. We repeat this grouping process until all coins have been used up.

Your task is to write a function that takes three input parameters: n (the total number of coins), g (the number of groups), and c (a constant) and returns an array containing the number of remaining coins at each grouping step. If n is 0, the function must return an empty list. For example, let $n = 10$, $g = 3$, and $c = 2$. The following transition show the remaining coins at each grouping step, as follows: 10, 3, 2 → 1; 6, 3, 2 → 0; 4, 3, 2 → 1; 2, 3, 2 → 2; 0, 3, 2. Therefore, [1, 0, 1, 2] represents the remaining coins in each step for $n = 10$, $g = 3$, and $c = 2$. In the first transition, 1 is obtained from (10 mod 3 = 1). In the second transition 6 is obtained from $10//3 * 2 = 6$, while 3 and 2 remain constant throughout the process. For some inputs, the expected outputs are illustrated in Table 3.9.

$$n//g * c \qquad (3.5)$$

Algorithm

The algorithm uses recursion to divide the coins into groups until no further groups can be formed. The following are the detailed steps involved in the algorithm:

1. Check if the input parameter n is equal to zero. If it is, return an empty list, as there are no coins available for splitting.
2. Calculate the number of coins that cannot constitute a complete group by computing the remainder when n is divided by g. Store this value in a variable named *remain*.

Table 3.9 The expected outputs for certain inputs to group coins

n, g c	Expected output
255, 128, 70	[127, 70]
301, 10, 1	[1, 0, 3]
49, 49, 2	[0, 2]
10, 2, 1	[0, 1, 0, 1]

3.10 Median of Triple Medians

3. Compute the total number of coins required to form all complete groups by multiplying the number of complete groups ($n//g$) by the size of each group (c). Save this value in a variable named *coins_needed*.
4. Recursively invoke the function with the input parameters *coins_needed*, g, and c until there are no more coins remaining to group.
5. Return an array containing the number of remaining coins at each grouping step.

The Python code for grouping the coins is depicted in Code 3.9.

Code 3.9 Python code for grouping the coins

```python
def Grouping_Coins(n, g, c):
    if n == 0:
        return []
    # how many coins cannot form a complete group
    remain = n % g
    
    '''
    (1) n //g:
    denotes groups of coins can be formed.
    (2) n //g * c:
    denotes coins are needed to form these groups.
    '''
    coins_needed = (n // g) * c
    remaining_coins = Grouping_Coins(coins_needed, g, c)
    return [remain] + remaining_coins
```

3.10 Median of Triple Medians

This challenge requires finding the median of a list of positive integers repeatedly until only a maximum of three numbers remain. Specifically, for each group of three numbers, their median must be computed and used as the input for the next iteration. This process continues until there are at most three numbers remaining, and their median is returned as the final result.

You are requested to write a Python function that takes a list of positive integers as input and returns the median of the last remaining numbers. For some inputs, the expected outputs are illustrated in Table 3.10.

Algorithm

To determine the median of medians, an algorithm is utilized that iteratively calculates the medians of sublists containing three elements within a specified list of positive integers until only three numbers are left. The process involves dividing the input list

Table 3.10 The expected outputs for certain inputs to find median of medians

Items	Expected output
[909, 4, 4, 7, 9, 12, 77, 45]	9
[1, 4, 90, 65, 3, 2]	4
[22, 40, 65, 80, 93, 21]	80
[100, 9, 12, 20, 3]	5

into groups of three items, computing their median, and appending it to the medians list. The function then recursively calls itself with the medians list until there are no more than three items remaining. When at most three items remain, the function sorts them in ascending order and returns the middle value as the ultimate median.

1. Take the length of the input list.
2. If the length of the input list is equal to 1, return that item as the median.
3. Otherwise, create an empty list called 'medians' to store the medians.
4. Split the input list into sublists of three items each using a for loop and range function with a step size of 3. Sort each sublist in ascending order and get the middle item as the median.
5. Append each median to the 'medians' list.
6. Recursively call the above step until there are no more than three items left.
7. When there are at most three items remaining, sort them in ascending order and get the middle item as the final median.
8. Return the final median as the output.

The Python code for finding the Median of medians (triple numbers) is depicted in Code 3.10.

Code 3.10 Python code for finding the Median of triple numbers

```
1  def insertion_sort(arr):
2      for i in range(1, len(arr)):
3          key = arr[i]
4          j = i - 1
5          while j >= 0 and key < arr[j]:
6              arr[j + 1] = arr[j]
7              j -= 1
8          arr[j + 1] = key
9      return arr
10
11 def median_of_medians(items):
12     # Take the length of the list
13     n = len(items)
14
15     '''
16         If there is only one item in the list,
```

```python
17              return that item as the median
18          '''
19      if n == 1:
20          return items[0]
21      else:
22          # Create an empty list to store medians
23          medians = []
24
25          '''
26          Split the list into sublists
27          of three items, sort them, and
28          get the middle item as the median
29          '''
30          for i in range(0, n, 3):
31              sublist = items[i:i+3]
32              if len(sublist) < 3:
33                  median = sublist[len(sublist)//2]
34              else:
35                  sorted_sublist = insertion_sort(sublist)
36                  median = sorted_sublist[1]
37              medians.append(median)
38
39          '''
40          Rursively call the function with
41          the medians list until there are
42          no more than three items left
43          '''
44          if len(medians) > 3:
45              return median_of_medians(medians)
46          else:
47              '''
48              If there are three or fewer items
49              left, sort them and get the middle
50              item as the final median
51              '''
52              sorted_items = insertion_sort(medians)
53              return sorted_items[len(sorted_items)//2]
```

3.11 Smallest Seven-Zero

In Western culture, the number seven holds significant importance and is often considered a symbol of good fortune. In contrast, the number zero is typically viewed as undesirable. However, when these two numbers are combined to form sequences

Table 3.11 The expected outputs for certain inputs to find smallest seven zero

n	Expected output
25	700
49	777777
1023	7777777777777777777777777777777
104	777777000

such as 7777777777777777 or 77700, known as 'seven-zero'. A remarkable theorem proves that for any positive integer n, there exist infinitely many integers of the seven-zero form that are divisible by n. The objective of this challenge is to identify the first seven-zero number for a given positive integer. It should be noted that having all seven zeros for a positive integer is not the desired outcome; rather, the aim is to find the first seven-zero (that is the smallest). Table 3.11 illustrates some examples of the input and output of this function.

Algorithm

In order to determine the smallest integer consisting solely of 7s or a combination of sevens and zeros for a given positive integer, an algorithm is employed that involves iterating over increasing powers of 10 until a multiple of 'n' comprising only of 7 (e.g. 777777) or sevens and zeros (e.g. 777777000) is discovered. The initial search space ranges from 1 to 10.

During each iteration, 7 is added to the product of the previous remainder and the power of 10 that corresponds to it. If a remainder appears more than once, this indicates that the seven-zero is a sequence of digits that constitutes a multiple of n which begins with 7 and ends with 0. If a remainder of 0 is reached, then a multiple of n beginning with the digit 7 has been found.

If no solution is identified within the current search range, the algorithm expands the search range and continues its iterations. The following are the detailed steps involved in the algorithm:

1. A dictionary named *remainder_positions* will be created to keep track of the remainders and their positions in each iteration. Initially, *current_remainder* is set to 0.
2. The initial values of *digits_min* and *digits_max* are set to 1 and 10 respectively. These values represent the search space for finding the first seven-zero, which is a multiple of n, and *num_digits* points from *digits_min* to *digits_max*.
3. Next, a while loop is started that runs until a seven-zero multiple of n is found.
4. Within the while loop, the values of *num_digits* are iterated over starting from *digits_min* and ending at *digits_max*.
5. For each value of *num_digits*, the next remainder is calculated using the formula $(current\_remainder * 10 + 7)\%n$.

3.11 Smallest Seven-Zero

6. If the current remainder is equal to zero, an integer consisting of a sequence of 7 is returned.
7. If the current remainder has appeared before, the previous block is utilized with the current block to form a multiple of n that begins with 7 and ends with 0. To do this, the position i of the previous occurrence of *current_remainder* in the *remainder_positions* dictionary is found and an integer consisting of sequence 7 followed by 0 using $'7' * (num_digits - i) +' 0' * i$ is returned.
8. If the current remainder has not appeared before, it is added to the *remainder_positions* dictionary with its position *num_digits*.
9. Once all iterations within the current search space are completed, the search space is increased by setting *digits_min* equal to *digits_max*, and multiplying *digits_max* by 10.
10. Steps 4–9 are repeated until a seven-zero multiple of n is found. If a seven-zero number is not identified, the iteration will continue indefinitely

The Python code for finding the smallest seven zero number is depicted in Code 3.11.

Code 3.11 Python code for seven zero

```python
def Smallest_Seven_Zero(n):
    # keep track of remainders and their positions
    remainder_positions = {}
    current_remainder = 0

    # set initial values for search space
    digits_min = 1
    digits_max = 10

    '''
    iterate over increasing powers of 10
    until we find a multiple of n
    '''
    while True:
      for num_digits in range(digits_min, digits_max):
          current_remainder = \
          (current_remainder*10 + 7) % n
          if current_remainder == 0:
              return int('7'*num_digits)
          elif current_remainder in remainder_positions:
              # combine previous block with current block
              i = remainder_positions[current_remainder]
              return int('7'*(num_digits-i) + '0'*i)
          else:
              remainder_positions[current_remainder] =\
              num_digits

      # increase search space if necessary
```

```
29          digits_min = digits_max
30          digits_max *= 10
```

3.12 Postfix Evaluate

In this task, the objective is to transform mathematical expressions written in postfix notation to infix notation, and subsequently, evaluate them. Infix notation refers to the mathematical expression where operators are written between pairs of operands, such as a + b. On the other hand, postfix notation refers to the mathematical expression where the operator is written after the operands, like ab+. For example, the given list is [2, 7, '+', 3, '*'] and its equivalent infix is $(2 + 7) * 3$, where the result is 27. This function must return the numerical result, not the equivalent infix. For some inputs, the expected outputs are illustrated in Table 3.12.

Algorithm

To evaluate the infix expression of the given postfix an algorithm is utilized that performs the following steps:

1. It iterates through each item in the given postfix expression.
2. If the item is an integer, it is appended to the Storage list.
3. If the item is an operator (+, −, *, /), the two most recent items (operands) are popped from the Storage list.
4. The corresponding operation (+, −, *, /) is then performed on these two operands, and the result is stored in the variable result, and result is then appended to the Storage list.
5. Steps 3–4 are repeated until all items in the postfix expression have been processed.
6. Finally, the result of the evaluation is returned by popping the last item from the Storage list.

The Python code for computing the result of converting postfix to infix and evaluating it is depicted in Code Code 3.12.

Table 3.12 The expected outputs for certain inputs to evaluate the infix of an postfix

Postfix	Expected output
[5, 6, '+', 7, '*']	77
[3, 7, 9, '*', '+']	66
[3, 7, 9, '/', '+']	3
[8, 2, '+']	10

3.12 Postfix Evaluate

Code 3.12 Python code for computing the result of postfix to infix

```python
1   def pop_last_item(input_list):
2       # Get and remove the last item from the input list
3       list_length = len(input_list)
4       last_item = input_list[list_length-1]
5       del input_list[list_length-1]
6       return last_item
7   def postfix_evaluate(Postfix):
8       '''
9       Create an empty list to
10      store operands and results
11      '''
12      Storage = []
13
14      '''
15      Iterate through each item
16      in the postfix expression
17      '''
18      for item in Postfix:
19          '''
20          If the item is an integer,
21          append it to the storage list
22          '''
23          if isinstance(item, int):
24              Storage.append(item)
25          else:
26              '''
27              If the item is an operator,
28              pop two most recent items from the storage list
29              '''
30              Number0, Number1 = \
31              pop_last_item(Storage), pop_last_item(Storage)
32              '''
33              Perform the corresponding operation on
34              the two numbers depending on the operator
35              '''
36              if item == '-':
37                  result = Number1 - Number0
38              elif item == '+':
39                  result = Number1 + Number0
40              elif item == '*':
41                  result = Number1 * Number0
42              elif item == '/':
43                  result = \
```

Table 3.13 The expected outputs for certain inputs to determine the stable state in Bulgarian solitaire

Points, k	Expected output
[3], 2	1
[3, 7, 8, 14], 4	0
[10, 10, 10, 10, 10, 5], 10	74
[3000, 2050], 100	0

```
44              Number1 // Number0 if Number0 != 0 else 0
45              Storage.append(result)
46
47      return pop_last_item(Storage)
```

3.13 Stable State in Bulgarian Solitaire

In this challenge, a list of n cards is given in the form of positive integers. The objective is to generate a new list of numbers that is different from the original list until a stable state is reached. A stable state is achieved when the elements of a list are the same as its preceding list, but the permutations can be different. There are two rules as follows: (1) The sum of the given list of numbers must be equal to a triangular number. The formula for the triangular number is provided in Eq. 3.6. (2) To create a new list in each step, one unit is subtracted from all the numbers, and the last element of the list is set equal to the length of the previous list. If the result of the subtraction is zero, it is removed. This process is repeated until a stable state is reached. Hence, the problem requires finding a list that satisfies both conditions. This process so-called is Bulgarian solitaire. For instance, if [3] be the given list of number and $k = 2$, counts the number of steps to reach a stable state. First, it must be checked if the triangular number is equal to the sum of the given list. To this end, $k \times (k+1)/2 \rightarrow 2 \times (2+1)/2 = 3$, so $3 = 3$. The following steps indicate the steps to reach a stable state: $[3] \rightarrow [2, 1] \rightarrow [1, 2, 0] \rightarrow [1, 2]$. As the elements of $[2, 1]$ and $[1, 2]$ are equal, it reaches a stable state, although $[2, 1]$ and $[1, 2]$ have different permutations. Therefore, after one step, it reached a stable state. Write a function in that as input, takes a list of n positive integers and a triangular number and returns the number of steps required to reach a stable state. Some examples of the input and output of this function on the Bulgarian solitaire problem are illustrated in Table 3.13.

$$\frac{k \times (k+1)}{2} \tag{3.6}$$

It is crucial to determine whether a stable state has been reached in this challenge. One possible technique for doing so involves sorting the current list and comparing

3.13 Stable State in Bulgarian Solitaire

it with the previous list. If the two lists are the same, it indicates that a stable state has been achieved. However, despite its seeming simplicity, this approach can be highly complex.

To determine the state (stable or not), two algorithms are employed: 'is_stable_state' and 'Stable State Bulgarian Solitaire'.

The 'Stable_State_in_Bulgarian_Solitaire' algorithm has the following steps:

1. It takes two arguments, *points* (a list of positive integers) and k (an integer). Initialize a variable named *number_of_moves* to 0.
2. Calculate the sum of points in the input list and store it in a variable named equation using the formula $k * (k + 1) // 2$.
3. Call the 'is_stable_state' algorithm that takes two arguments, points (a list of positive integers) and k (an integer).
4. Check whether the sum of the points list equals equation. If not, return zero to indicate that the input list is invalid. Loop until the stable state is achieved:

   - Subtract 1 from each element in the points list.
   - Remove any zero values in the points list.
   - Append the length of the current points list to the end of the points list.
   - Increment the *number_of_moves* by 1.
   - Check whether the current state is stable using the 'is_stable_state' algorithm.

5. Return the final value of *number_of_moves*.

The 'is_stable_state' algorithm has the following steps:

1. It takes two arguments, *points* (a list of positive integers) and k (an integer).
2. Create an array named *count* with $k + 1$ zeros inside the function.
3. Loop through each value p in the points list and increment the corresponding element in count if p is less than or equal to k.
4. Loop through every integer i from 1 to k, and check whether the *count[i]* equals 1. If not, return False indicating that the state is unstable.
5. If the loop completes without returning False, return True indicating that the state is stable.

The python code for computing the stable state in the Bulgarian solitaire is depicted in Code 3.13.

Code 3.13 Python code for computing the the Bulgarian solitaire

```
1  def Stable_State_in_Bulgarian_Solitaire(points, k):
2      # Initialize the number of moves to zero
3      number_of_moves = 0
4
5      '''
6      Equation for calculating the
7      sum of points
8      '''
9      equation = k * (k + 1) // 2
```

```python
def is_stable_state(points, k):
    """
    To keep track of the frequency of each integer
    between 1 and k present in the list of points,
    we use a counting array named 'count'. If any
    integer appears multiple times or not at all,
    the stable state has not yet been achieved,
    and as a result, the function returns False.
    Conversely, if every integer between 1 and k
    appears exactly once, the function returns True,
    indicating that a stable state has been reached.
    """
    count = [0] * (k+1)
    for p in points:
        if p <= k:
            count[p] += 1
    for i in range(1, k+1):
        if count[i] != 1:
            return False
    return True

'''
Check if the input points satisfy
the condition for Bulgarian solitaire
'''
if sum(points) == equation:
    # Loop until the stable state is reached
    while is_stable_state(points, k)!=True:
        # Decrease the size of each pile by 1
        for i in range(len(points)):
            points[i] -= 1

        points_Length = len(points)
        # Remove piles with zero size
        points = [p for p in points if p > 0]

        '''
        Add a new pile with the
        size of the removed piles
        '''
        points.append(points_Length)

        # Increment the number of moves
        number_of_moves += 1
```

```
55
56        # Return the total number of moves
57        return number_of_moves
```

3.14 Computing the Rectangular Towers in Manhattan Skyline

There is a list of rectangular towers as (s, e, h), where s, e, and h are a start, end, and height in the x-coordinate, respectively. Write a python function that as input, takes tuple (s, e, h), and as an output, returns the area of the rectangular towers. Notice that the common area of the towers should not be computed twice. For some inputs, the expected outputs are illustrated in Table 3.14.

Algorithm

To compute the area of the given towers, two algorithms are employed: 'divide' and 'Manhattan_Skyline'. The algorithm 'divide' has the following steps:

1. It takes two parameters, *start* and *end*. If *start* equals *end*, return a list containing two elements: [*towers*[*start*][0], *towers*[*start*][2]] and [*towers*[*start*][1], 0].
2. Otherwise, calculate the midpoint between *start* and *end*, and recursively call the left and right halves of the towers list. Merge the results from the left and right halves into a new list using the following steps:

   - Compute the mid position
   - Initialize an empty list called 'merged', and two variables i and j to 0.
   - Set $h1$ and $h2$ variables to None.
   - Loop over every element in the merged list by iterating over the range of len(left) + len(right).

Table 3.14 The expected outputs for certain inputs to compute the rectangular towers in Manhattan skyline

Towers	Expected output
[(2, 6, 98), (1, 0, 9)]	383
[(3, 7, 1)]	4
[(−8, 6, 3), (6, 14, 11), (0, 4, −5)]	130
[(2, 550, 222), (1, 0, 4)]	121652

- If i is greater than or equal to the length of left, append a list containing [right[j][0], right[j][1] if $h1$ is None else max(h1, right[j][1])] to merged, increment j by 1, and continue the loop.
- If j is greater than or equal to the length of right, append a list containing [left[i][0], left[i][1] if $h2$ is None else max(h2, left[i][1])] to merged, increment i by 1, and continue the loop.
- If the x-coordinate of the left element at index i is less than or equal to the x-coordinate of the right element at index j, append a list containing [left[i][0], left[i][1] if $h2$ is None else max(h2, left[i][1])] to merged, set $h1$ to the y-coordinate of the left element at index i, and increment i by 1.
- Otherwise, append [right[j][0], right[j][1]] to merged if $h1$ is None else max(h1, right[j][1])] to merged, set $h2$ to the y-coordinate of the right element at index j, and increment j by 1.

3. Return the merged list.

The algorithm 'Manhattan_Skyline' has the following steps:

1. It takes *towers* as input.
2. Compute the area enclosed by the skyline using the 'divide'.
3. Return the computed area.

The python code for computing the area of the rectangular towers is depicted in Code 3.14.

Code 3.14 Python code for computing the active area of the rectangular towers

```
def rectangular_towers_in_manhattan_skyline(towers):

    ''' divide_and_conquer computes the
    skyline for a given range of towers
    '''
    def divide_and_conquer(start, end):
        '''
        If there is only one tower in the range,
        return its two corners
        '''
        if start == end:
            return [[towers[start][0],
                     towers[start][2]],
                    [towers[start][1], 0]]

        '''
        Divide the range into two halves and
        recursively compute the skylines for each half
        '''
        mid = (start + end) // 2
        left = divide_and_conquer(start, mid)
```

3.14 Computing the Rectangular Towers in Manhattan Skyline

```
22          right = divide_and_conquer(mid+1, end)
23
24          '''
25          Merge the two skylines using
26          a merge sort-like approach
27          '''
28          merged = []
29          i, j = 0, 0
30          # Heights of the left and right towers
31          h1, h2 = None, None
32
33          for x in range(len(left) + len(right)):
34
35              if i >= len(left):
36                  merged.append([right[j][0], right[j][1]
37                      if h1 is None else max(h1, right[j][1])])
38                  j += 1
39
40              elif j >= len(right):
41                  merged.append([left[i][0], left[i][1]
42                      if h2 is None else max(h2, left[i][1])])
43                  i += 1
44                  '''
45                  If the next element in the left
46                  skyline is to the left of the next
47                  element in the right skyline, append
48                  the left element and update its height
49                  '''
50
51              elif left[i][0] <= right[j][0]:
52                  merged.append([left[i][0], left[i][1]
53                      if h2 is None else max(h2, left[i][1])])
54                  h1 = left[i][1]
55                  i += 1
56                  '''If the next element in the right
57                  skyline is to the left of the next element
58                  in the left skyline, append the right
59                  element and update its height
60                  '''
61              else:
62                  merged.append([right[j][0], right[j][1]
63                      if h1 is None else max(h1, right[j][1])])
64                  h2 = right[j][1]
65                  j += 1
66
```

```
67          # Return the merged skyline
68          return merged
69
70      # Get the length of the towers list
71      n = len(towers)
72
73      # Compute the skyline for the entire range of towers
74      result = divide_and_conquer(0, n - 1)
75
76      '''
77      Compute the area enclosed by the
78      skyline using the trapezoidal rule
79      '''
80      area = sum((result[i+1][0]-result[i][0]) *
81              result[i][1] for i in range(len(result)-1))
82
83      # Return the computed area
84      return area
```

3.15 Cut Rectangular into Squares

A tuple (a, b) represents the length and width of a rectangle, respectively. In this challenge, write a Python function that takes a tuple (a, b) as input and returns the minimum number of steps required to convert the rectangle into a square.

The expected inputs and respective outputs are provided in Table 3.15.

Algorithm

The best approach for solving this problem is the recursive function. The aim of the algorithm is to determine the minimum number of cuts required to divide a given rectangular area into equal-sized squares. To achieve this, the algorithm begins by checking if the given rectangle is already a square. If that is the case, no cuts are

Table 3.15 The expected outputs for certain inputs to cut rectangular into squares

a, b	Expected output
(7, 7)	0
(17, 10)	6
(5, 3)	3
(7, 9)	5

3.15 Cut Rectangular into Squares

required, and the algorithm returns 0. However, if the rectangle is not a square, the algorithm recursively divides it into smaller rectangles until they become squares. During this process, memoization is used to store previously computed results and avoid redundant computations. Once all the rectangles have been transformed into squares, the algorithm compares the results obtained by cutting the final square both horizontally and vertically. It then selects the solution with the minimum number of cuts as the optimal solution, memoizes the result and returns it.

1. It takes 'a' (length), 'b' (width) as input.
2. Create a dictionary called 'memo' to store values.
3. Check if both 'a' and 'b' are equal. If they are, return 0 as there are no squares to cut.
4. Check if either 'a' or 'b' equals 1. If so, return the maximum of 'a' and 'b' minus 1 since there can only be one square in that case.
5. Create a key by making a tuple of ('a', 'b') to check if the key exists in the memo dictionary. If it does, return the corresponding value.
6. Initialize answer as ('a' − 1) * 'b'.
7. Loop through variable 'i', from 1 to 'a' // 2 + 1.
8. Calculate the minimum between answer and ('a' − 'i', 'b', memo) plus ('i', 'b', memo) plus 1.
9. Loop through variable 'i', from 1 to 'b' // 2 + 1.
10. Calculate the minimum between answer and ('a', 'b' − 'i', memo) plus ('a', 'i', memo) plus 1.
11. Add the key-value pair to the memo dictionary.
12. Return answer.

The Python code for computing the minimum number of states wanted to reach a square is depicted in Code 3.15.

Code 3.15 Python code for computing the minimum number of states wanted to reach a squares

```python
def RectToSquares(a, b, memo={}):
    '''
    Check if the given rectangles
    are already squares
    '''
    if a == b:
        return 0

    if a == 1 or b == 1:
        return (max(a, b) - 1)

    key = (a, b)
    '''Check if we have already computed
    the answer for this input
    '''
    if key in memo:
```

```
17            return memo[key]
18
19
20
21        '''
22        Initialize the answer as the
23        maximum possible value
24
25        Represents the maximum possible
26        number of cuts required to cut
27        the rectangle into squares.
28        '''
29        answer = (a - 1) * b
30
31        # Cutting the rectangle horizontally
32        for i in range(1, a // 2 + 1):
33            '''
34            We recursively compute the
35            minimum number of cuts
36            required for the two smaller rectangles
37            '''
38            # formed by the horizontal cut
39            horizontal_cut = RectToSquares(a - i, b, memo)
40            vertical_cut = RectToSquares(i, b, memo)
41            # Add 1 to account for the initial cut
42            total_cuts = horizontal_cut + vertical_cut + 1
43            # Update the answer if we find a better solution
44            answer = min(answer, total_cuts)
45
46        # Cutting the rectangle vertically
47        for i in range(1, b // 2 + 1):
48            '''
49            Recursively compute the minimum
50            number of cuts required for
51            the two smaller rectangles
52            '''
53            # formed by the vertical cut
54            horizontal_cut = RectToSquares(a, i, memo)
55            vertical_cut = RectToSquares(a, b - i, memo)
56            # Add 1 to account for the initial cut
57            total_cuts = horizontal_cut + vertical_cut + 1
58            # Update the answer if we find a better solution
59            answer = min(answer, total_cuts)
60
61        # Memoize the answer
```

3.16 Eliminating Corners

Table 3.16 The expected outputs for certain inputs to eliminate corners

Points	Expected output
[(3, 3), (3, 8), (8, 3)]	1
[(0, 1), (4, 5), (3, 2)]	0
[(5, 0), (1, 3), (1, 4), (2, 0), (2, 2), (2, 3), (4, 0), (4, 0)]	2

```
62      memo[key] = answer
63      return answer
```

3.16 Eliminating Corners

For a given set of points on a two-dimensional integer grid, a 'corner' refers to three points in the exact form of (x, y), $(x, y + h)$, and $(x + h, y)$ for some $h > 0$, representing the tip and two wings of the corner. Given a list of points sorted by their x-coordinates, with ties resolved by their y-coordinates, this function should return the minimum number of points that need to be removed from the list to eliminate all corners. The expected inputs and respective outputs are provided in Table 3.16.

Algorithm

To eliminate all corners formed by three points in a given list of points, while minimizing the number of removed corners, the algorithm first identifies all possible corners that can be formed by three points in the input list using nested loops. It then utilizes recursion with memoization to determine the minimum number of corners that must be removed to eliminate all corners from the list. The Python code for counting the minimum number of points to remove all corners is depicted in Code 3.16.

Code 3.16 Python code for counting the minimum number of points to eliminate all corners

```
 1  '''
 2  Recursively remove the minimum number of corners
 3  '''
 4  def MinCornerRemover(corners, memo):
 5      '''
 6      Base case: if the list of corners is empty,
 7      no removals are needed
 8      '''
 9      if len(corners) == 0:
10          return 0
```

```python
11
12      '''
13      Check if the current list of corners
14      has been processed before
15      '''
16      key = tuple(corners)
17      if key in memo:
18          return memo[key]
19
20
21      removals = len(corners)
22
23      '''
24      Initialize a set to keep track of points
25      that have already been processed
26      '''
27      points_set = set()
28      # Loop through each corner in the list
29      for corner in corners:
30          '''
31          Loop through each point
32          in the current corner
33          '''
34          for point in corner:
35              '''
36              If the point has not been processed
37              yet, add it to the set
38              '''
39              if point not in points_set:
40               points_set.add(point)
41               '''
42               Compute the remaining corners
43               by excluding the current point
44               '''
45               remaining_corners = \
46               [c for c in corners if point not in c]
47               '''
48                  Recursively compute the minimum number
49                  of removals for the remaining corners
50               '''
51               '''and add 1 to account for the
52                  current point being removed
53               '''
54               removals = \
55               min(1 + MinCornerRemover
```

3.16 Eliminating Corners

```
56                 (remaining_corners, memo), removals)
57
58         '''
59         Memoize the result for the
60         current list of corners
61         '''
62         memo[key] = removals
63         # Return the minimum number of removals
64         return removals
65
66  '''
67  Eiminating all corners from
68   the given list of points
69  '''
70  # It takes list of tuples representing points as input
71  def cut_corners(points):
72      # Initialize an empty list to store the corners
73      corners = []
74      # Loop through each point in the list
75      for i in range(len(points)):
76          x = points[i][0]
77          y = points[i][1]
78          '''
79          Check if there is another point
80          with the same y-coordinate
81          '''
82          # that is to the right of the current point
83          for j in range(i+1, len(points)):
84              if points[j][1] == y:
85                  h = points[j][0] - x
86                  '''Check if there is a point h units
87                  above the current point
88                  '''
89                  '''and to the right of the other
90                  point with the same y-coordinate
91                  '''
92                  if h > 0 and (x, y+h) in points:
93                      '''
94                      Append the tuple representing the
95                      corner to the list of corners
96                      '''
97                      corners.append\
98                      (((x, y), (x+h, y), (x, y+h)))
99
100     '''
```

```
101        Initialize an empty dictionary to
102        store previously computed results
103        '''
104    memo = {}
105        '''
106        Call the function to recursively remove
107        the minimum number of corners using memoization
108        '''
109    return MinCornerRemover(corners, memo)
```

3.17 Leibniz Triangle

In Pascal's triangle, each number is equal to the sum of the two numbers directly above it, whereas in Leibniz's triangle, each number is equal to the sum of its two bottom numbers. It is required to write a function that takes the leftmost row values as input and returns the corresponding value of the last row at the desired position. The expected outputs for some inputs are presented in Table 3.17.

For instance, two numerical examples of Leibniz's triangle are depicted in Fig. 3.2. The inputs for the first examples (A) are 2 and 3, and the position is $range(2)$. The numbers in the $range$ function start from 0, so the positions of zero and one return 2 and 1, respectively.

Table 3.17 The expected outputs for certain inputs for Leibniz triangle

LeftMost_Values,Positions	Expected output
[1, 2, 3], range(3)	[3, −1, 0]
[4, 7, 9, 3], range(4)	[3, 6, −8, 7]
[88, 90, 1, 0, 9], range(4)	[9, −9, 10, 78]
[20, 95], range(2)	[95, −75]

3.17 Leibniz Triangle

Fig. 3.2 Two numerical example for Leibniz's triangle

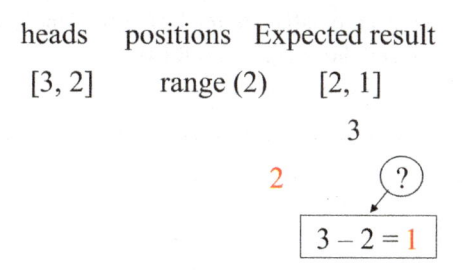

heads   positions Expected result

[3, 2]   range (2)   [2, 1]

$$3$$
$$2 \quad\; ?$$
$$3 - 2 = 1$$

A)   **Leibniz numerical example**

heads   positions Expected result

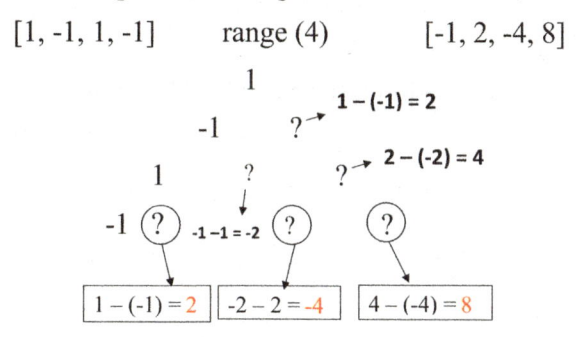

[1, -1, 1, -1]   range (4)   [-1, 2, -4, 8]

$$1$$
$$-1 \quad\; ?\; \rightarrow 1-(-1)=2$$
$$1 \quad\; ? \quad\; ? \rightarrow 2-(-2)=4$$
$$-1\; ? \quad -1-1=-2 \;\; ? \qquad\; ?$$

| 1 − (−1) = 2 | −2 − 2 = −4 | 4 − (−4) = 8 |

B)   **Leibniz numerical example**

Algorithm

To solve the given task, we utilize an algorithm that consists of the following steps:

1. It takes two inputs: *LeftMost_Rows*, a list of values representing the leftmost row of a Leibniz triangle, and *positions*, a list of integers representing positions in the last row of the same triangle for which we want to find the values.
2. Initialize an empty list to store the result for each wanted position.
3. Initialize a dictionary to store intermediate values computed during the algorithm.
4. Compute the values in the first column of the triangle using the provided leftmost row values.
5. Store these values in the memoization table with their corresponding indices as keys.
6. Compute the remaining values in the triangle using the values from the previous row.
7. Each value in the triangle is computed as the difference between the sum of the two bottom values in the current row and the value directly above it in the previous row.
8. Store these values in the memoization table with their corresponding indices as keys.

9. Retrieve the desired values in the last row of the triangle from the memoization table and append them to the result list.
10. Return the list of results.

The Python code for Leibniz's triangle is depicted in Code 3.17.

Code 3.17 Python code for Leibniz's triangle

```
'''
This function takes two arguments:
a list of values representing
the leftmost row of a Leibniz triangle
 and a list of integers representing
positions in the last row of the same triangle
for which we want to find the values.
'''
def Leibniz_triangle(LeftMost_Rows, positions):
    '''
    Initialize an empty list to store
    the result for each position
    '''
    result = []
    '''
    Initialize a dictionary to store intermediate
    values computed during the algorithm
    '''
    memo = {}

    '''
    Compute the values in the first column of the
    triangle using the provided leftmost row values.
    '''
    for i in range(len(LeftMost_Rows)):
        memo[(i, 0)] = LeftMost_Rows[i]

    '''
    Compute the remaining values in the triangle
    using the values from the previous row.
    '''
    for i in range(1, len(LeftMost_Rows)):
        for j in range(1, i+1):
            '''
            Each value in the triangle is computed
            as the difference between the sum
            of the two bottom values
            '''
            value = memo[(i-1, j-1)] - memo[(i, j-1)]
```

```
40                    memo[(i, j)] = value
41
42        '''
43        Retrieve the desired values in the
44        last row of the triangle
45        from the memoization table and
46        append them to the result list.
47        '''
48        for i in positions:
49            result.append(memo[(len(LeftMost_Rows)-1, i)])
50
51        # Return the list of results.
52        return result
```

3.18 Collatzy Distance

In this challenge, a positive integer n is given, and it is aimed at reach to some positive integers by applying $3*n+1$ and $n//2$, where the formula is taken from the collatz conjecture. Let there be two *start* and *goal* numbers with h layers each layer contains some positive integer numbers, and the first layer is initialized with *start* number. All positive integers have the same distance in each layer. It starts from *start* number and generates the numbers in each layer once by $3*n+1$ and once by $n//2$ until reaches *goal* number. For example, consider the following example. Let $start = 10$ and $goal = 20$, in each step, it generates some positive integers with the same distance, until reaches the *start* number. As the first layer is initialized with *start*, 10 is considered.

- Layer 1: Once $3*n+1$ for 10 is applied and once $n//2$ for 10 is applied. $(3n+1) = (3*10+1) = 31, (n//2) = (10//2) = 5$
- Layer 2: Once $3*n+1$ for 31 and 5 is applied and once $n//2$ for 31 and 5 is applied. $(3*31+1) = 94, 16, (3*5+1) = 16, (31//2) = 15, (5//2) = 2$.
- Layer 3: 283, 47, 46, 7, 49, 8,1
- Layer 4: 850,141, 142, 23, 139, 22, 3, 148, 24, 25, 4, 0
- Layer 5: 2551, 425, 424, 70, 71, 11, 69, 67, 10, 74, 73, 12, 75, 13,...
- Layer 6: 7654, 1275, 1276, 212, 1273, 211, 35, 34, 31, 37, 6, 40,...
- Layer 7: 22963, 3827, 3826, 637, 3829, **20**,...

It can be seen that in layer 7, it found the *goal* number of **20**. Write a function that as input, takes *start* and *goal* and returns the layer number that visits *goal* number. For some inputs, the expected outputs are illustrated in Table 3.18.

Table 3.18 The expected outputs for certain inputs for Collatzy distance

Start, goal	Expected output
20, 321	9
22, 15	8
4, 7	2
10, 20	2

Algorithm

To achieve the goal number for the stated challenges, a breadth-first search (BFS-based) technique is employed. The BFS algorithm traverses all the nodes at a particular level before moving on to explore the subsequent level. In this instance, each generated layer resulting from the application of the Collatz conjecture formulas denotes a level of exploration. The algorithm initiates its exploration from the start_num, which represents the first layer, and generates successive layers by applying the same formulas. The following steps are involved in the algorithm:

1. It takes *start_num* and *goal_num* (positive integers).
2. Initialize the variables *num_layers* and *curr_layer*, where *num_layers* = 0 and *curr_layer* is a set containing only the starting number, i.e., start_num.
3. While the *goal_num* is not in *curr_layer*, do the following:

    - Create an empty set called *next_layer* (As we dont need duplicated numbers).
    - For each number in *curr_layer*, compute the the following ones:
    - *num*//2.
    - 3 ∗ *num* + 1.
    - Add them to *next_layer*.
    - Update *curr_layer* to be equal to *next_layer*.
    - Increment *num_layers* by 1.

4. Return *num_layers* when *goal_num* is found in *curr_layer*.

The Python code for computing collatzy distance is depicted in Code 3.18.

Code 3.18 Python code for computing collatzy distance

```
1  def get_collatz_distance(start_num, goal_num):
2      """
3      Takes a start number and a goal number
4      and generates all layers of positive
5      integers between them using the Collatz
6      sequence to find the shortest path from
7      start_num to goal_num.
8      """
9      # initialize the layer counter
10     num_layers = 0
```

3.18 Collatzy Distance

```python
    '''
    initialize the current layer as a
    set containing only the start_num
    '''
    curr_layer = {start_num}
    '''
    keep looping until goal_num
    is in the current layer
    '''
    while goal_num not in curr_layer:
        '''
        initialize the next layer
        as an empty set
        '''
        next_layer = set()

        for num in curr_layer:
            '''generate the next numbers in
            the Collatz sequence for each number
            in the current layer
            '''
            next_layer.update([
                num // 2,
                    3 * num + 1
            ])
        '''
        set the current layer to be the
        newly generated layer
        '''
        curr_layer = next_layer
        '''
        increment the layer counter
        '''
        num_layers += 1
    '''
    return the number of layers
    it took to reach goal_num
    '''
    return num_layers
```

120　　3　Math

Table 3.19 The expected outputs for certain inputs for the problem of sum of two squares

n	Expected output
50	(7,1)
145	(12, 1)
1480	(38, 6)
540	None

3.19 Sum of Two Squares

In this challenge, you will be given a positive integer 'n' and your task is to calculate the squares of two numbers that add up to 'n'. For example, if 'n' is equal to 145, the answer would be (12,1). If there are multiple answers, the answer with the highest maximum number should be selected. For example, when 'n' is equal to 50, there are two possible answers: (7,1) and (5,5). As 7 is greater than 5, the answer should be (7,1). You are required to write a function that takes a positive integer 'n' as input and returns the two numbers whose squares add up to 'n'. If there are no such numbers, the function should return 'None'. The expected inputs and outputs for some test cases are illustrated in Table 3.19.

Algorithm

The algorithm finds the bases by first setting a maximum value for the first base, then checking all candidate values for the second base. If a candidate value for the second base leads to a pair of bases that satisfy the conditions, the function returns this pair. The following provides an outline of the algorithm's steps.

1. It takes a positive integer n.
2. Calculate the maximum value of the first base as the square root of half of n.
3. Create a set of candidate values for the second base from 1 to the maximum value of the first base.
4. Loop through the candidate values of the second base.
5. For each candidate value of the second base, calculate the difference between n and the square of the second base.
6. Calculate the square root of the difference.
7. Check if the square of the root equals the difference.
8. If it does, return the pair of bases whose squares sum up to n.
9. If no such pair exists, return None.

The Python code for computing the sum of two squares is depicted in Code 3.19.

Code 3.19 Python code for computing the sum of two squares

```
1  def get_squares_summing_to_n(n):
2      """
```

```
3    This function returns a tuple of two
4    integers whose squares sum up to `n`.
5    """
6    # The maximum value of first base
7    max_first_base = int((n/2)**0.5)
8    # Set of candidate values for the second base
9    '''
10   the second base cannot be less than the first base,
11   so 1 + max_first_base
12   '''
13   candidates = set(range(1, 1 + max_first_base))
14   # Loop through the candidate values of second base
15   for second_base in candidates:
16       '''
17       Calculate the difference between n
18       and square of second base
19       '''
20       difference = n - second_base**2
21       # Calculate the square root of the difference
22       root = int(difference ** 0.5)
23       '''
24       Check if square of the root
25       equals the difference
26       '''
27       if root ** 2 == difference:
28           '''
29           Return the pair of bases
30           whose squares sum up to `n`
31           '''
32           return (root, second_base)
33   # If no such pair exists, return None
34   return None
```

3.20 Has Three Summer

In this challenge, a sorted list of positive integers must be evaluated to determine if there exist three numbers whose sum is exactly equal to a given goal number. It is required to write a function that takes as input a sorted list of positive integers, and returns true if three numbers are found whose sum matches the goal number, otherwise it returns false. The expected outputs for certain inputs are presented in Table 3.20.

Table 3.20 The expected outputs for certain inputs for three summer problem

ListNumbers, goal	Expected output
[3, 5, 6, 8, 9, 21], 14	True
[2, 4, 8, 16, 32], 16	False
[550, 600, 2000, 3000, 4000], 900	False
[1, 2, 16, 79, 80, 340], 83	True

Algorithm

There is a solution that employs three nested for loops to traverse all the numbers and compare their sum with the target number. Although this method is simple but has high time complexity. To reach the outlined task, an algorithm is utilized that includes following steps.

1. It takes a sorted list of numbers and a target number as input.
2. Check if the length of the input list is less than 3. If it is, return False.
3. Iterate through the sorted list one by one.
4. For each number in the list, compute the new target number by subtracting the current number from the goal number.
5. In the next step, the algorithm utilizes the two summer algorithm to reach the answer. Two summer takes a list of numbers and a target number as input, and returns True if any two numbers in the list add up to the target number. Otherwise, it returns False.
6. The two summer algorithm has the following steps:
    a. Check if the length of the input list is less than 2. If it is, return False.
    b. Initialize two pointers, one at the start of the list and the other at the end of the list.
    c. While the start pointer is less than the end pointer:
       1. Compute the sum of the two numbers at the current positions of the start and end pointers.
       2. If the sum equals the target number, return True.
       3. If the sum is less than the target number, increment the start pointer to move to a larger number.
       4. If the sum is greater than the target number, decrement the end pointer to move to a smaller number.
    d. If no match was found, return False.
7. If such a pair exists, return True.
8. If no match was found for any number in the list, return False.

The Python code for computing three summers is depicted in Code 3.20.

Code 3.20 Python code for computing three-summers

```
1  def has_two_sum(numbers, target):
```

3.20 Has Three Summer

```python
    '''
    If the length of the list
    is less than 2, it is not
    possible to find two numbers
    that add up to target
    '''
    if len(numbers) < 2:
        return False

    '''Initialize pointers at the
    start and end of the list
    '''
    start = 0
    end = len(numbers) - 1

    '''Keep moving the pointers until
    they meet in the middle
    '''
    while start < end:
        two_sum = numbers[start] + numbers[end]

        '''If the sum of the two numbers
        equals target, we have found a match
        '''
        if two_sum == target:
            return True

            '''If the sum is less than target,
        increment the start pointer to
        move to a larger number
            '''
        elif two_sum < target:
            start += 1

        # If the sum is greater than target,
        #decrement the end pointer to move
        #to a smaller number
        else:
            end -= 1

    '''If no match was found,
    return False
    '''
    return False
```

```python
'''
The main function that checks
if there exist three numbers in
a sorted list that add up
 to a target number
'''
def has_three_summers(numbers_list, goal):
    '''If the length of the list is
    less than 3, it is not possible
    to find three numbers that
    add up to the target
    '''
    if len(numbers_list) < 3:
        return False

    # Iterate through the list one by one
    for i in range(len(numbers_list)):
        '''Get the current number
        from the list
        '''
        x = numbers_list[i]
        '''Compute the new goal
        by subtracting
        '''
        new_goal = goal - x
        
        '''
        Check if there exist two numbers
        in the remaining list that add
        up to the new target
        '''
        if has_two_sum(numbers_list[i+1:], new_goal):
            return True

    '''
    If no match was found,
    return False
    '''
    return False
```

3.21 Perfect Power

A positive integer n is considered a perfect power if it can be expressed in the form $base^{power}$, where $base$ and $power$ are both integers greater than 1. Write a function that as input takes a positive integer n, and if find $base^{power} = n$, return true, else return false. For instance, for $n = 64$, there is $base = 2$ and $power = 6$, $2^6 = 64$. For some inputs, the expected outputs are illustrated in Table 3.21.

Algorithm

To determine whether a positive integer n is a perfect power or not, an algorithm is used that performs a binary search to find the base of the perfect power. The algorithm starts by setting the power to 2 and continues looping until it either finds a perfect power or determines that there is no perfect power less than or equal to n. This algorithm has the following steps as follows.

1. It takes a positive integer n.
2. Set *power* to 2.
3. Loop while True:
4. If $2**power > n$, return False.
5. Set *low* to 2 and *high* to n.
6. While *low* is less than or equal to *high*:
   1. Set *mid* to the floor division of $(low + high)$ by 2.
   2. Set *guess* to *mid* raised to the *power*.
   3. If *guess* is equal to n, return True.
   4. Else if *guess* is less than n, set low to $mid + 1$.
   5. Else, set *high* to $mid - 1$.
7. If $base^{power}$ is equal to n, return True
8. Increment power by 1

The Python code for checking the perfect powers of the positive integers is depicted in Code 3.21.

Table 3.21 The expected outputs for certain inputs for determining if a number is a perfect power

n	Expected output
2	False
27	True
729	True
1369	True

Code 3.21 Python code for checking the perfect powers of the positive integers

```python
def perfect_power(n):
    '''
    In Python, ** is used for the power operator
    '''
    power = 2

    while True:
        if 2 ** power > n:
            return False

        # Find largest power using binary search approach
        low = 2
        high = n
        while low <= high:
            mid = (low + high) // 2
            guess = mid ** power

            if guess == n:
                return True
            elif guess < n:
                low = mid + 1
            else:
                high = mid - 1

        # Increment power
        power += 1
```

3.22 Lunar Multiplication

In the lunar multiplication problem, the multiplication of two numbers is equal to their minimum, and their sum is equal to their maximum. For example, $2+7=7+2=7$, $2 \times 7 = 7 \times 2 = 2$ since the lunar operations have commutative properties. Write a function that takes two n-digit integers a and b and returns their lunar multiplication as output. The same as the ordinary addition, the unit element for addition is zero, while for any natural number n, 9 is the unit element, where $2 \times 9 = 2=, 7 \times 9 = 7$, $9 \times 63 = 63$. Consider the following example. Let $a = 10$ and $b = 21$, digit by digit the multiplication is done.

- (1×10) $\to$ $1 \times 0 = 0$) $\to$ ($1 \times 1 = 1$) $\to$ 10
- (2×10) $\to$ $2 \times 0 = 0$) $\to$ ($2 \times 1 = 1$) $\to$ 10.
- $+\frac{10}{10} = 110$.

3.22 Lunar Multiplication

Table 3.22 The expected outputs for certain inputs for lunar multiplication

a, b	Expected output
10, 21	110
170, 76	1760
45, 96	455
8, 7	7

Some examples of the input and expected output of the lunar multiplication problem are illustrated in Table 3.22.

Algorithm

The algorithm for this problem is straightforward and simple. In one step, it performs the multiplication between each two number, and then it adds the multiplied numbers with considering their carries. The Python code for lunar multiplication is depicted in Code 3.22.

Code 3.22 Python code for lunar multiplication

```python
def multiply_by_lunar(a, b):
    '''
    A function to add two
    numbers in lunar addition
    '''
    def add_in_lunar(x, y):
        # Convert x and y to strings, reverse them
        '''
        Because the calculation from
        least significant digit is specified.
        '''
        x_reversed = rev(x)
        y_reversed = rev(y)

        # Initialize sum as an empty string
        sum_reversed = ''

        '''Calculate maximum length of numbers
        and perform addition
        '''
        max_len = max(len(x_reversed), len(y_reversed))

        i = 0
        while i < max_len:
```

```python
            '''Perform lunar addition of digits at
            same positions in reversed strings
            '''
            sum_reversed += \
            max(x_reversed[i:i+1], y_reversed[i:i+1])
            i += 1


        '''
    Reverse the result and convert it back
    to integer before returning
    '''
        return int(sum_reversed[::-1])

    # Convert a and b to strings, reverse them
    '''
    addition on digits at corresponding positions
    starting from the least significant digit.
    '''
    a_reversed = rev(a)
    b_reversed = rev(b)

    # Initialize numbers array and answer variable
    numbers = []
    answer = 0

    i = 0
    j = 0
    d = ''
    while i < len(b_reversed):
        if j < len(a_reversed):
            '''
            Append minimum digit of a and b at
            current positions in reversed strings
            '''
            d += min(a_reversed[j], b_reversed[i])
            j += 1
        else:
            '''If we reach end of a, append
            remaining digits of b to d
            '''
            numbers.append('0' * i + d)
            i += 1
            j = 0
            d = ''
```

```
70        if d:
71            # Append remaining digits of d to numbers
72            numbers.append('0' * i + d)
73
74        '''Add numbers of each step using
75        the add_in_lunar function
76        '''
77        for num in numbers:
78            answer = add_in_lunar(answer, num[::-1])
79
80        return answer
81
82
83  # A function to reverse a number
84  def rev(x):
85      x_reversed=''
86      for i in range(len(str(x)) - 1, -1, -1):
87          x_reversed += str(x)[i]
88      return x_reversed
```

3.23 *n*-th Term of Recaman Sequence

In the field of computer science, the Recamán sequence is a type of sequence that exhibits a recurrence relation, whereby the computation of a new element is dependent on the preceding elements. To illustrate, in order to derive the *n*-th term of the Recamán sequence (where $n = 1$), the following steps are taken: It should be noted that the elements of this sequence are determined by the recursive formula provided in Eq. 3.7. The zeroth term, a_0, is always equal to zero. To calculate a_1, we find that $0 - 1 = -1 < 0$, thereby rejecting the second condition and proceeding with the third condition, which gives $a_1 = 0 + 1$. Write a function that takes a positive integer *n* as input and returns the *n*-th term of the Recamán sequence.

$$x = \begin{cases} 0 & \text{if } n == 0 \\ a_{n-1} - 1 & \text{if } a_{n-1} - 1 > 0 \text{ and the number is not already in sequence} \\ a_{n-1} + n & \text{else} \end{cases}$$

(3.7)

For some inputs, the expected outputs are illustrated in Table 3.23.

Table 3.23 The expected outputs for certain inputs for finding n-th term of Recaman sequence

n	Expected output
17	25
83	72
919	756
632	308

Algorithm

To find the n-th term, the algorithm initializes the sequence with the first element, and a set to keep track of used elements. It then iterates over the remaining elements up to 'n', and for each element it calculates the next element using Eq. 3.7. If the next element is negative or already used, it uses the alternative formula to calculate the next element. The next element is then added to the sequence, and marked as used in the set. Finally, the algorithm returns the n-th element of the sequence. The Python code for computing the n-th term of the recaman sequence is depicted in Code 3.23.

Code 3.23 Python code for computing the n-th term of the recaman sequence

```python
def Nth_term_of_recaman_sequence(n):
    '''
    Returns the nth item in the Recamán sequence
    '''
    # Initialize the sequence with the first element
    seq = [0]
    # Initialize a set to keep track of used elements
    seen = {0}
    # Iterate over the remaining elements up to n
    for i in range(1, n+1):
        # Get the previous element of the sequence
        prev = seq[-1]
        '''Calculate the next element
        using the following formula
        '''
        next = prev - i
        '''
        If the next element is negative
        or already used, the alternative formula
        '''
        if next < 0 or next in seen:
            next = prev + i
        # Add the next element to the sequence
        seq.append(next)
        # Mark the next element as used in the set
```

```
26            seen.add(next)
27        # Return the nth element of the sequence
28        return seq[-1]
```

3.24 Van Eck Sequence

Consider a sequence S that is made up of positive integers. The first term of S is 0, is there a zero number before the current zero? no, so the sequence is 0, 0. Again, for the last number (second zero), this question is asked, is there a zero before the second zero? yes, so the sequence is 0, 0, 1. Again, for the last number (one) this question is asked, is there a 'one' before the one number? no, so the sequence is 0, 0, 1, 0. Again, for the last number (0) this question is asked, is there a 'zero' before the zero number? yes, so the sequence is 0, 0, 1, 0, 2, and so on. In a simple statement, it will count the number of steps to reach the first number asked, where from the tail of the sequence is started to find the number asked, and if see it, ends the counting. This popular sequence is the van eck sequence. It is an integer sequence, and recursive-based. It is recursive because the next term is calculated based on the previous terms. In this challenge, the task is to compute and return the n-th term of the van eck sequence. Write a function that as input, takes a positive integer n, and returns the n-th term of the van eck sequence. For some inputs, the expected outputs are illustrated in Table 3.24.

Algorithm

The algorithm to determine the n-th term of the Van Eck sequence is presented below:
1. Initialize the sequence with the first term, which is 0.
2. Create an empty dictionary to store the index of each value in the sequence.
3. For each index i from 0 to $n - 1$

Table 3.24 The expected outputs for certain inputs for finding n-th term of Van Eck sequence

n	Expected output
258	3
25	4
2	1
2089	0

1. Get the previous value in the sequence, which is sequence[i].
2. Check if the previous value has occurred before by looking it up in the dictionary.
3. If it has occurred before, calculate the difference between the current index and the previous index stored in the dictionary for the previous value.
4. Append the difference to the sequence.
5. If the previous value hasn't occurred before, append 0 to the sequence.
6. Update the dictionary with the current index i for the previous value.

4. Return the n-th term in the sequence, which is $sequence[n-1]$.

The Python code for computing the n-th term of the van eck sequence is depicted in Code 3.24.

Code 3.24 Python code for computing the n-th term of the van eck sequence

```python
"""
Returns the nth term
in the Van Eck sequence.
"""
def find_nth_term_in_van_eck_sequence(n):
    '''
    Initialize the sequence
    with the first term
    '''
    sequence = [0]
    '''
    Create a dictionary to store
    the index of each value in the sequence
    '''
    value_to_latest_index = {}
    for i in range(n):
        # Check if the previous value has occurred before
        previous_value = sequence[i]
        if previous_value in value_to_latest_index:
            '''
            If it has, calculate the difference between
            the current index and the previous index
            '''
            difference = i - \
            value_to_latest_index[previous_value]
            # Add the difference to the sequence
            sequence.append(difference)
        else:
            # If it hasn't, add 0 to the sequence
            sequence.append(0)
        '''
```

3.25 Non-consecutive Fibonacci Numbers

Table 3.25 The expected outputs for certain inputs for finding non-consecutive Fibonacci numbers

n	Expected output
53	[34, 13, 5, 1]
25	[21, 3, 1]
31	[21, 8, 2]
3009	[2584, 377, 34, 13, 1]

```
32              Update the index of the current
33              value in the dictionary
34              '''
35              value_to_latest_index[previous_value] = i
36      # Return the last value in the sequence
37      return sequence[-1]
```

3.25 Non-consecutive Fibonacci Numbers

According to Zeckendorf's theorem, any arbitrary positive integer can be expressed as a sum of distinct non-consecutive Fibonacci numbers in a unique manner. Write a function that takes a positive integer n as input and returns a list of non-consecutive Fibonacci numbers which add up to n. The list must be in descending order. The expected outputs for certain inputs are illustrated in Table 3.25.

Algorithm

The algorithm involves generating a sequence of numbers using the Fibonacci sequence, sorting the generated sequence in descending order using the insertion sort algorithm, and selecting specific elements from the sorted sequence that meet certain criteria. The following steps indicate how the algorithm works.

1. It takes positive integer n.
2. Initialize the variables previous_fibonacci_number to 0, current_fibonacci_number to 1, fibonacci_numbers to a list containing previous_fibonacci_number and current_fibonacci_number, selected_fibonacci_numbers to an empty list, and current_sum to the sum of previous_fibonacci_number and current_fibonacci_number.
3. Use a while loop to generate Fibonacci numbers up to n as follows:
    1. Append current_sum to fibonacci_numbers.
    2. Update previous_fibonacci_number to current_fibonacci_number.
    3. Update current_fibonacci_number to current_fibonacci_number + previous_fibonacci_number.

4. Update current_sum to the sum of previous_fibonacci_number and current_fibonacci_number.
5. Continue the loop until current_sum is greater than n.
6. Sort fibonacci_numbers in descending order.

4. Use a for loop to select non-consecutive Fibonacci numbers that sum to n as follows:
5. Initialize the variable sum_of_fibonacci_numbers to 0.
6. For each number in fibonacci_numbers, do the following:
7. If sum_of_fibonacci_numbers + number is less than or equal to n, append number to selected_fibonacci_numbers, update sum_of_fibonacci_numbers to sum _of _fibonacci_numbers + number.
8. If sum_of_fibonacci_numbers is equal to n, return selected_fibonacci_numbers.

The Python code for the Fibonacci sum is depicted in Code 3.25.

Code 3.25 Python code for fibonacci sum

```
1  def sort_desc(arr):
2      for i in range(1, len(arr)):
3          curr = arr[i]
4          j = i - 1
5          while j >= 0 and curr > arr[j]:
6              arr[j + 1] = arr[j]
7              j -= 1
8          arr[j + 1] = curr
9      return arr
10
11 def get_non_consecutive_fibonacci_numbers_summing_to_n(n):
12     prev = 0
13     curr = 1
14     fib_nums = [prev, curr]
15     selected_fib_nums = []
16     curr_sum = prev + curr
17
18     while curr_sum <= n:
19         fib_nums.append(curr_sum)
20         prev, curr = curr, curr + prev
21         curr_sum = prev + curr
22
23     fib_nums = sort_desc(fib_nums)
24
25     sum_of_fib_nums = 0
26     for num in fib_nums:
27         if sum_of_fib_nums + num <= n:
28             selected_fib_nums.append(num)
```

```
29          sum_of_fib_nums += num
30          if sum_of_fib_nums == n:
31              return selected_fib_nums
```

3.26 Fibonacci Word

The Fibonacci word is a sequence that resembles the ordinary Fibonacci sequence, with the difference that instead of the sum of the two previous numbers, their concatenation is considered. The first term of this sequence is 0, and the second term is 01 (as a string). Each subsequent term is obtained by concatenating the two previous terms. For example, the third term is obtained by concatenating the first and second terms, which results in 010. Similarly, the fourth term is obtained by concatenating the second and third terms, which results in 01001, and so on. This challenge requires a function that takes an input integer k and returns the k-th character of the Fibonacci word, where the sequence is a string consisting of the characters 0 and 1. The function must be implemented to handle the desired inputs and provide the expected outputs as illustrated in Table 3.26.

Algorithm

A straightforward approach to resolving the challenge is to produce the sequence up to its k-th position and extract said element. However, this method results in heightened time and space complexity. To find the k-th character, the equations Eqs. 3.8 and 3.9 are used, with phi denoting the golden number.

$$phi = \frac{1+\sqrt{5}}{2} \qquad (3.8)$$

$$\lfloor (k+2) * phi/(1+(phi*2)) \rfloor - (k+1) * phi/(1+(phi*2)) \qquad (3.9)$$

Table 3.26 The expected outputs for certain inputs for finding k-th term in Fibonacci word

k	Expected output
65	0
170	0
98	1
2022	1

The Python code for Fibonacci word is depicted in Code 3.26.

Code 3.26 Python code for fibonacci word

```
1   import decimal
2   def sqrt(num):
3       decimal.getcontext().prec = 165
4       return decimal.Decimal(num).sqrt()
5   def floor(x):
6       return int(x - 1) if x < 0 else int(x)
7   def fibonacci_word(k):
8       # Taking the fifth root
9       root_5 = sqrt(5)
10
11      # Calculating the golden number
12      phi = (1 + root_5) / 2
13      x= floor((k + 2) * phi / (1 + (phi * 2)))
14      y=floor((k + 1) * phi / (1 + (phi * 2)))
15      # Returning the kth character
16      return x-y
```

3.27 Finding the Most Point Line

A point on the two-dimensional integer grid is represented by a two-tuple of x- and y-coordinates, such as (2,5) and (10,3). When two distinct points are present in the plane, they define exactly one line that passes through both of them. However, this unique line extends infinitely in both directions and intersects an infinite number of other points on the same plane, although proving this proposition from first principles may prove challenging. Write a function that as input, takes a list of points on the grid, and returns the line that has the largest number of points from the given list. This function must not return the line itself, but only count the number of points on the considered line. For some inputs, the expected outputs are illustrated in Table 3.27.

Table 3.27 The expected outputs for certain inputs for finding the most point line on the two-dimensional integer grid

k	Expected output
[(4, 4), (6, 6), (3, 2), (1, 4)]	2
[(14, 4), (0, 0), (1, 2), (3, 0)]	2
[(3, 15), (1, 4), (2, 6), (8, 7), (3, 8)]	3

3.27 Finding the Most Point Line

Algorithm

The algorithm takes a list of 2D integer grid points as input and returns the maximum number of points on a line passing through of the points. The algorithm first checks if the length of the input list is less than 3, in which case it returns the length of the list. Otherwise, it iterates through each pair of distinct points in the list, calculates the slope of the line passing through those points, and keeps track of the frequency of each slope using a dictionary. It also keeps track of the number of duplicate points. Finally, it returns the maximum number of points on a line by adding the current maximum number of points to the number of duplicates and comparing it to the previous maximum. The Python code for the line with the most points is depicted in Code 3.27.

Code 3.27 Python code for line with most points

```python
def calculate_gcd(a, b):
    '''
    Compute the greatest common
    divisor of two integers a and b
    '''
    while b != 0:
        a, b = b, a % b
    return a

def count_points_on_line(points):
    n = len(points)
    '''
    If there are less than three points,
    they always form a line.
    '''
    if n < 3:
        return n

    max_points_on_line = 0
    i = 0
    '''
    Iterate through each point
    in the list of points
    '''
    while i < n:
        '''
        Create a dictionary to keep track
        of slopes and their frequency
        '''
        slope_count = {}
        num_duplicates = 1
```

```python
        cur_max_points_on_line = 0
        j = i+1
        # Compare with other points to calculate the slope
        while j < n:
            if points[i] == points[j]:
                num_duplicates += 1
            else:
                dx = points[j][0] - points[i][0]
                dy = points[j][1] - points[i][1]
                if dx == 0:
                    slope = float('inf')
                else:
                    gcd = calculate_gcd(dy, dx)
                    slope = (dy//gcd, dx//gcd)

                '''
                Add the slope to the dictionary
                or increment its frequency
                '''
                slope_count[slope] = \
                slope_count.get(slope, 0) + 1

                '''
                Update the maximum number of points
                on a line with the current slope
                '''
                cur_max_points_on_line = \
                max(cur_max_points_on_line,
                slope_count[slope])

            j += 1

        '''
        Update the overall maximum number of points
        on a line considering current point
        '''
        max_points_on_line =\
        max(max_points_on_line,
        cur_max_points_on_line + num_duplicates)
        i += 1

    return max_points_on_line
```

3.28 Is a Balanced Centrifuge

Scientists use centrifuges to test tubes, but it is essential to balance the tubes in the centrifuge to avoid disrupting its performance or causing serious damage due to its high speed. The balancing can be achieved by placing the tubes in a manner such that they either face each other or form a regular pattern. For example, in Fig. 3.3, part A shows a centrifuge with six different holes, while part B shows four tubes placed in four of the holes. As can be seen in part B, the test tubes are balanced, with the two tubes on the left and right facing each other. For certain input values, the expected outputs are listed in Table 3.28.

The objective of this challenge is to determine whether a given situation is balanced, given the values of n and k. A function is required that takes as input the number of holes and tubes, representing n and k respectively, and returns true if the situation is balanced, or false otherwise.

Algorithm

Given n and k, the first step involves determining the prime factors of n and storing them in a list L. Subsequently, it is necessary to verify whether it is feasible to obtain both $n - k$ and k using the prime factors in L. For example, consider the case where

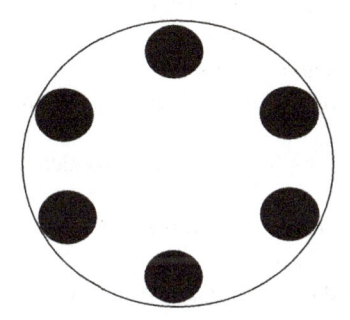

A) Centrifuge with six holes, n=6

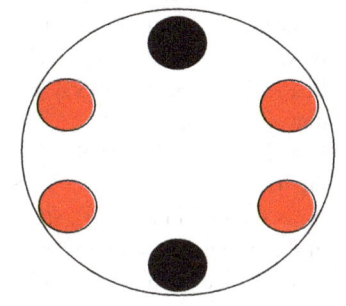

B) Centrifuge with six holes and 4 tube (red ones), n=6, k=4

Fig. 3.3 Balanced centrifuges with example

Table 3.28 The expected outputs for certain inputs determining if a centrifuge is balanced or not

n, k	Expected output
76, 19	True
23, 3	False
113, 18	False
60, 2	True

$n = 6$ and $k = 3$. The list of prime factors for n is [2, 3]. As both k and $n - k$ can be expressed using the prime factors in L. Specifically, $k = 3$ is included in L, and $n - k = 6 - 3 = 3$ is also present in L, indicating a balanced situation for $n = 6$ and $k = 3$. An additional example involves the values $n = 15$ and $k = 8$, whereby the list of prime factors is [5, 3]. While it is possible to obtain k by adding 5 and 3, it is not possible to obtain $n - k$, which in this case is 7, using the numbers in the list. Consequently, the function returns false.

To solve the problem, two algorithms are being considered. The first algorithm is used to determine the prime factors of n, while the second algorithm is used to verify whether n and $n - k$ can be created using the prime factors obtained from the first algorithm.

Finding Prime Factors

An empty list to store prime factors is created. The first prime number is equal to 2. As long as n is greater than 1, n is divided by 2. If the remainder is zero and the number was not added to the considered list, we add it to the list. In each step, n is updated with the quotient. If n is no longer divisible by 2, the next prime is considered, and this process is repeated until $n < 1$.

It takes a positive integer 'n' as input and returns a list of prime factors of 'n'. The step-by-step explanation of the algorithm is outlined below.

1. An empty list called 'primes' is initialized to store the prime factors.
2. A variable 'i' is initialized to 2, which is the first prime number.
3. A while loop is used to check if the number 'n' is greater than 1.
4. Inside the loop, it checks if 'n' is divisible by 'i'. If 'n' is divisible by 'i', it means that 'i' is a factor of 'n'.
5. If 'i' is a prime factor of 'n' and is not already in the 'primes' list, it is added to the list.
6. 'n' is then divided by 'i', so that we can continue to check if 'i' is a prime factor of the remaining value of 'n'.
7. If 'i' is not a prime factor of 'n', 'i' is incremented by 1 and the loop continues to the next iteration.
8. The loop continues until 'n' is less than or equal to 1.
9. Finally, the list of 'primes' containing all the prime factors of 'n' is returned as the output of the algorithm.

Coin Change

The coin change algorithm is used to determine whether a given number can be expressed as a combination of values from a specified list. This algorithm takes as input a number and an array, and checks whether the number can be formed using elements of the given list. The idea to solve this problem is to use dynamic

3.28 Is a Balanced Centrifuge

programming. Specifically, the algorithm creates a list with $n + 1$ elements, and initializes it to zero, except for the first element ($ways[0]$), which is set to 1. This represents the fact that there is one way to make a total of 0 using no coins from the list. The i-th element of this array is equal to the number of times that the number i can be made with the numbers of the array. Hence, the last element of the array indicates the number of ways the number n can be made by the array elements. The Python code for the balanced centrifuge is depicted in Code 3.28.

Code 3.28 Python code for balanced centrifuge

```
1  def is_balanced_centrifuge(n, k):
2      def get_primes(n):
3          """
4          Returns a list of prime factors
5          for the given number.
6          """
7          primes = []
8          i = 2
9          # Iterate over all possible factors of the number
10         while n > 1:
11             if n % i == 0:
12                 # Add factor to the list of primes
13                 if i not in primes:
14                     primes.append(i)
15                 '''Divide the number by the factor
16                 to get the next factor
17                 '''
18                 n /= i
19             else:
20                 '''
21                 If it is not a divisor,
22                 try the next number
23                 '''
24                 i += 1
25         return primes
26  
27      def can_build_number(arr, n):
28          """
29          Determines whether a number can be made
30          with the elements of a given list.
31          """
32          '''
33          Create a list of ways to make
34          each number up to the target number
35          '''
36         ways = [0] * (n + 1)
```

```
        # There is only one way to make 0
        ways[0] = 1
        # Iterate over all elements in the list
        for i in range(len(arr)):
            # Iterate over all possible numbers to build
            for j in range(arr[i], n + 1):
                '''
                Add the number of ways to make the current
                number using the current element
                '''
                ways[j] += ways[j - arr[i]]
    '''
    If there is at least one way to make
    the target number, return True
    '''
    return ways[n] != 0

# Get the list of prime factors for the n
primes = get_primes(n)
# Balancing is checked
# can be made using the prime factors
return can_build_number(primes, k)\
    and can_build_number(primes, n - k)
```

Chapter 4
Number

This chapter talks about 21 number-based problems. These challenges are explained with some examples and then programmed in Python. The problems are listed below:

1. Checking if a number is Cyclop
2. Checking if there is a domino cycle in a list of numbers
3. Extracting increasing digits from a given string
4. Expanding integer intervals
5. Collapsing integer intervals
6. Checking if a dice is left-handed or right-handed
7. Bonus to repeated numbers
8. Nearest first smaller number
9. First preceding object by k smaller numbers
10. Finding n-th term of calkin wilf
11. Reverse ascending subarrays
12. Smallest integer powers
13. Sorting cycles of a graph
14. Obtaining Numbers in Balanced Ternary System
15. Is strictly ascending
16. Priority Sorting
17. Sort positives, keep negative numbers
18. Sorting numbers first, characters second
19. Sorting dates
20. Sorting by alphabetical order and length
21. Sorting by digit count.

4.1 Cyclop Numbers

A non-negative integer is referred to as a cyclops number if it meets the following conditions: the number of digits is odd, the middle digit (also known as the 'eye') is zero, and all other digits in the number are non-zero. To verify whether an non-negative integer input is a cyclops number or not, a Python function needs to be developed. The function should take a non-negative integer as input and check if it satisfies the aforementioned conditions. If the input integer is a cyclops number, the function should return True; else, it should return False (refer to Table 4.1 for expected outputs for certain inputs).

To check if a number is a cyclops number or not, the algorithm follows these steps:

1. It takes an integer n as input.
2. Convert the integer n into a string of digits and store it in the variable digits.
3. Check if the length of digits is even. If it is, return False as a cyclops number has an odd number of digits.
4. Find the middle index of digits by dividing the length of digits by 2 using integer division (//) and store it in the variable $middle\_number$.
5. Check if the digit at the index $middle\_number$ in digits is equal to 0. If it is not, return False as a cyclops number must have a 0 in the middle.
6. Count the number of zeroes in digits and store the result in the variable count. If the count is greater than 1, return False as a cyclops number can only have one 0. Otherwise, return True as the input n satisfies all the conditions of a cyclops number.

The Python code for cyclop number is depicted in Code 4.1.

Code 4.1 Python code for cyclop number problem

```
1  def cyclop_number(n):
2      digits = str(n)
3      if len(digits) % 2 == 0:
4          return False
5
6      middle_number = len(digits) // 2
7      if digits[middle_number] != '0':
8          return False
9
```

Table 4.1 The expected outputs for certain inputs for checking if a number is a cyclop or not

n	Expected output
11000	True
709	True
11318	False
6022	False

```
10        count = digits.count('0')
11        if count > 1:
12            return False
13
14    return True
```

4.2 Is a Domino Cycle

A domino tile can be represented by a two-tuple (x, y), where x and y are positive integers. The objective of this challenge is to identify a domino cycle from a given set of tiles. For example, $[(5, 2), (2, 3), (3, 4), (4, 5)]$ has a domino cycle. The first domino tile $(5, 2)$ is a connection with $(2, 3)$; $2 \rightarrow 2$. The domino tile $(2, 3)$ is a connection with $(3, 4)$; $3 \rightarrow 3$, and the other are: $4 \rightarrow 4, 5 \rightarrow 5$. Write a function that as input, takes a list of dominoes, and returns true if there is a domino cycle, otherwise returns false. For some inputs, the expected outputs are illustrated in Table 4.21.

Algorithm

The algorithm must check if y and x of the two compared tuples are not equal in each step. If they are not equal, it must return False; otherwise, it must return True. The Python code for determining the domino cycle is depicted in Code 4.2.

Code 4.2 Python code for determining the domino cycle

```
1  def Is_a_domino_cycle(tiles):
2      index = 0
3      while index < len(tiles):
4          # The comparing condition
5          if tiles[index][0] != tiles[(index-1)%len(tiles)][1]:
6              '''
7              If the condition is not met only once,
8              false is returned.
```

Table 4.2 The expected outputs for certain inputs for checking if there is a domino cycle

Tiles	Expected output
[(5, 2), (2, 3), (3, 4), (4, 5)]	True
[(3, 4), (4, 2), (2, 3), (3, 1), (4, 2)]	False
[(3, 4), (4, 2), (2, 3), (3, 1), (4, 2), (2, 4), (6, 3), (3, 2)]	False
[(6, 4), (4, 5), (5, 6)]	True

```
 9                '''
10                    return False
11            index += 1
12     return True
```

4.3 Extract Increasing Digits

In this challenge, a string of digits is given and it is guaranteed that the digits are 0, 1, 2, 3, 4, 5, 6, 7, 8, 9, and the task is to find the increasing digits. For instance, consider 457990, which has the increasing digits 4, 5, 7, 9 and 90. Write a function that as input, takes a string containing digits, and returns the list of increasing digits. For some inputs, the expected outputs are illustrated in Table 4.3.

Algorithm

The algorithm in two blocks checks the increasing digits. At first, an empty string *concatnumbers* is defined, and it is set $b = 1$. If $b = 1$, it enters the first block, and if $b = 0$, it enters the second block. For each digit of the given string, in the first block, with the linear search if the next number was greater than the previous number and $b = 1$, then the next number is appended into an array *storage* and updates the previous number. In the second block, if the comparison was not met in the first block, then $b = 0$, and make concatenation for the current number and its next, and put it in *concatnumbers*. If *concatnumbers* was greater than the previous number, then store it into *storage*, and make empty *concatnumbers*. As long as a concatenation of the next numbers is less than or equal to the previous number the concatenation is repeated for the next numbers. When $b = 0$ it never can enter the first block, because no longer single digits are greater than previous numbers, and entering the first block is meaningless. The Python code for extracting the increasing digits is depicted in Code 4.3.

Table 4.3 The expected outputs for certain inputs for extracting increasing digits

Digits	Expected output
'457990'	[4, 5, 7, 9, 90]
'1'	[1]
'13900456'	[1, 3, 9, 45]
'27811700'	[2, 7, 8, 11, 70]

4.3 Extract Increasing Digits

Code 4.3 Python code for extracting the increasing digits

```python
def extract_increasing_digits(digits):
    number=[]
    concatnumbers=''
    for x in str(digits):
        number.append(int(x))
    # Previous number
    prev= number[0]
    storage=[]
    # If there is just one digit
    storage=[number[0]]
    number=number[1:]
    b=1
    for next_ in number:
        '''
        This block extracts the increasing digits
        that are single
        such as 2,3,4
        '''
        if next_ > prev and b==1:
            storage.append(next_)
            prev=next_
        else:
            '''
            This block extracts the increasing digits
            that have more than one digit
            such as 90, 990, 9990
            '''
            b=0
            concatnumbers +=str(next_)
            if int(concatnumbers)>prev:
                storage.append(int(concatnumbers))
                prev=int(concatnumbers)
                '''
                Makes emepty concatnumbers until forms
                another number with the digits
                more than previous one
                '''
                concatnumbers=''
    return storage
```

4.4 Expand Integer Intervals

A range of consecutive positive integers can be represented as a string that includes the first and last values, separated by a dash (-). For example, '1,17-21,43,44' is a valid interval. The task at hand is to expand this interval into a list of individual numbers. To accomplish this, you are requested to write a function that takes a string of digits as input and returns a comma-separated list of numbers. It should be noted that the digits in the input string are guaranteed to be arranged in ascending order. The expected inputs and corresponding outputs are illustrated in Table 4.4.

Algorithm

At first, the algorithm must split the string by ','. In the next step, it must separate the single numbers from the intervals. The single numbers are stored in a list, and using a for loop iterates through the start and end of the interval, and each time the included numbers into the range are stored in a list. Outlined below are the detailed steps involved in the algorithm

1. It takes the string *intervals* as input.
2. Check if *intervals* is empty. If it is, return an empty list.
3. Create an empty list called *result* to store the expanded integers.
4. Split the input string into individual intervals using the comma as the delimiter and store them in a list called *intervals_list*.
5. For each interval in *intervals_list*, check if it contains only one number (i.e., if there is no dash '-'). If so, convert the number to an integer using int() and append it to the *result* list.
6. If the interval contains a range of numbers (i.e., if there is a dash '-'), split the interval by the dash and convert the *start* and *end* of the range to integers using *int*().
7. Generate all the numbers between the *start* and *end* using the range() function and add them to the result list using the extend() method.
8. Once all the intervals have been processed, return the final *result* list containing all the expanded integers.

Table 4.4 The expected outputs for certain inputs to expand integer intervals

Intervals	Expected output
'1, 7 − 16, 120 − 124, 568'	[1, 7, 8, 9, 10, 11, 12, 13, 14, 15, 16, 120, 121, 122, 123, 124, 568]
'2 − 10'	[2, 3, 4, 5, 6, 7, 8, 9, 10]
'2, 3 − 5, 19'	[2, 3, 4, 5, 19]
17	[17]

4.4 Expand Integer Intervals

The Python code for expanding the intervals is depicted in Code 4.4.

Code 4.4 Python code for expanding the integer intervals

```
 1   def Expanding_Integer_Intervals(intervals):
 2       '''
 3       If there are no given numbers,
 4       then returns an empty list
 5       '''
 6       if intervals == '':
 7           return []
 8       '''
 9       create an empty list to store
10       the expanded integers
11       '''
12       result = []
13       '''
14       split the input string
15       into individual intervals
16       '''
17       intervals_list = intervals.split(',')
18
19       for interval in intervals_list:
20           '''
21           Check if the interval
22           only contains one number
23           '''
24           if '-' not in interval:
25               '''
26               convert it to an integer
27               and add to the result list
28               '''
29               result.append(int(interval))
30
31           else:
32               # If the interval contains a range of numbers
33               # extract the start and end of the range
34               start_end = interval.split('-')
35               start = int(start_end[0])
36               end = int(start_end[1])
37               '''
38               generate all numbers between start and
39               end, and add to result list
40               '''
41               '''
42               range(x,y)---range(x,y-1),
```

```
43                    so (x,y+1) to consider the last item
44            '''
45                    result.extend(range(start, end+1))
46         '''
47         Return the final list containing
48         all the expanded integers
49         '''
50         return result
```

4.5 Collapse Integer Intervals

This function is the inverse of the previous problem. Write a function that as input takes a list of numbers and returns a string of digits and intervals in the form 'first-last' such that are comma separated. For some inputs, the expected outputs are illustrated in Table 4.5.

Algorithm

The algorithm to solve this challenge is rule-based. In fact, it iterates over a list of integers and identifying ranges of consecutive integers.

1. It takes a list of positive integer numbers as input.
2. Initialize two variables called 'start' and 'end' to None.
3. For each integer 'item' in the input list of integers 'items', do the following:

    a. If 'start' is None, set both 'start' and 'end' to 'item'.
    b. If 'item' is equal to 'end + 1', update 'end' to 'item'.
    c. If 'item' is not adjacent to the previous 'end', add the current range to 'ranges', where If 'start' is equal to 'end', append 'str(start)' to the 'ranges' list.
    d. Otherwise, append 'str(start)' + '−' + 'str(end)' to the 'ranges' list.
    e. Then, start a new range by setting both 'start' and 'end' to 'item'.

Table 4.5 The expected outputs for certain inputs to collapse integer intervals

Items	Expected output
[1, 7, 8, 9, 10, 11, 12, 13, 14, 15, 16, 120, 121, 122, 123, 124, 568]	'1, 7 − 16, 120 − 124, 568'
[2, 3, 4, 5, 6, 7, 8, 9, 10]	2 − 10
[2, 3, 4, 5, 19]	'2 − 5, 19'
[17]	17

4. If 'start' is not None, add the final range to 'ranges' using the same logic as the previous step.
5. Join the ranges with commas and return the resulting string.

The Python code for collapsing the numbers is depicted in Code 4.5.

Code 4.5 Python code for collapsing the numbers

```python
def collapse_integer_intervals(items):
    ranges = []
    start = end = None

    for item in items:
        if start is None:
            # Start a new range
            start = end = item
        elif item == end + 1:
            # Expand the current range
            end = item
        else:
            # Add the current range to the list
            if start == end:
                ranges.append(str(start))
            else:
                ranges.append(str(start) + '-' + str(end))

            # Start a new range
            start = end = item

    # Add the final range to the list
    if start is not None:
        if start == end:
            ranges.append(str(start))
        else:
            ranges.append(str(start) + '-' + str(end))

    # Join the ranges with commas
    return ",".join(ranges)
```

4.6 Left-Handed Dice

Consider a dice that have six sides with pips from one to six such that, there are eight corners, and in each corner, there are three visible numbers. From left or right, the dice have a different view. For instance, consider Fig. 4.1, part b, from the left-

Fig. 4.1 Example of the right-handed and left-handed

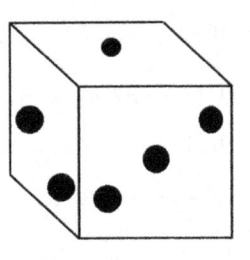

(a) Right-handed

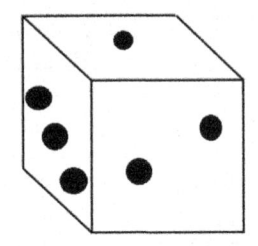

(b) Left-handed

Table 4.6 The expected outputs for certain inputs to determine if a dice is left-handed

n	Expected output
(4, 2, 1)	True
(6, 3, 2)	True
(6, 5, 4)	False

handed (clockwise) the reading is done each time. For the first time, 1, 2, 3 is read; the next number after 1 is 2, so reading from 2 is considered, so 2, 3, 1; the next number after 2 is 3, so reading from 3 is considered, so 3, 1, 2. Therefore, with views of left-handed for the corner 1, 2, 3, there are three permutations of 1, 2, 3, 2, 3, 1, and 3, 1, 2. There are eight different corners for left-handed and each with three numbers, so $8 * 3 = 24$. Likewise, the same is true for right-handed. In total there are 24+24= 48 permutations in left-handed and right-handed. Write a Python function that as input, takes a corner and determines whether is left-handed or right-handed; if it is left-handed, returns true, otherwise, returns false. For some inputs, the expected outputs are illustrated in Table 4.6.

Algorithm

The permutations of left-handed ones are small, it is enough to find 24 left-handed corners, and if the input matches the considered corner, it returns true, false otherwise.

4.7 Bonus to Repeated Numbers

The Python code for determining the left-handed or right-handed is depicted in Code 4.6.

Code 4.6 Python code for determining the left-handed or right-handed

```
1  def is_left_handed_dice(pips):
2      Left_Handed = [(1, 2, 3), (3, 1, 2), (2, 3, 1),
3                     (1, 4, 2), (2, 1, 4), (6, 3, 2),
4                     (4, 2, 1), (5, 6, 4), (4, 5, 6), (6, 4, 5),
5                     (3, 6, 5), (5, 3, 6),
6                     (6, 5, 3), (5, 4, 1), (1, 5, 4), (4, 1, 5),
7                     (2, 4, 6), (6, 2, 4),
8                     (4, 6, 2), (1, 3, 5), (5, 1, 3), (3, 5, 1),
9                     (2, 6, 3), (3, 2, 6)]
10
11     if pips in Left_Handed:
12         return True
13     else:
14         return False
```

4.7 Bonus to Repeated Numbers

Suppose an individual is walking on the street and comes across two taxis. One has the number 1729, while the other has the number 6666. It is evident that the number 6666 is more easily remembered due to the repeating sequence of digits. The human mind tends to recall numbers that have a repeating sequence more quickly. For instance, people tend to remember hours like 12 : 12 faster than ones without a repeating sequence, such as 10 : 23. This challenge aims to draw attention to sequences of numbers that have repeated digits. A score is assigned based on the number of repeated digits in the sequence. If the sequence has two repeated digits, the score is one. If it has three repeated digits, the score is ten. If it has four repeated digits, the score is one hundred, and so on. Furthermore, if the sequence ends with a repeating digit, the score is doubled. For example, 12333 is a number that has repeated numbers, and its lowest number ends with a repeated number. The scoring measure is represented by Eq. 4.1.

$$score = \begin{cases} 10^{(k-2)} & if\ there\ is\ any\ repeated\ sequence \\ 2 \times 10^{(k-2)} & if\ the\ lowest\ number\ is\ in\ repeated\ sequence \end{cases} \quad (4.1)$$

Write a function that as input, takes a positive integer n, and if visits a repeated sequence score it with $10^{(k-2)}$ each time, or if n ends with the repeating sequence, score it with $2 \times 10^{(k-2)}$. For some inputs, the expected outputs are illustrated in Table 4.7.

Table 4.7 The expected outputs for certain inputs to give a bonus to repeated numbers

n	Expected output
1233	2
1569	0
17777	200
588885515555	301

Algorithm

The algorithm takes integer n and iterates over each digit in the input number, and checks whether the current digit is the same as the previous digit. If it is the same, the algorithm increments a counter for the number of repeated digits. If it is different, the algorithm scores the count of repeated digits so far and resets the counter for the new digit. Finally, the algorithm scores any remaining repeated digits at the end of the number. The algorithm has the following steps.

1. It takes an integer n.
2. Initialize a variable called *score* to zero.
3. Initialize a variable called *count* to one.
4. Initialize a variable called $prev\_digit$ to None.
5. Convert the integer n to a string and iterate over each digit in the string:
    a. If the current digit is the same as the previous digit ($prev\_digit$), increment the count variable.
    b. If the current digit is different from the previous digit:
    c. If *count* is greater than 1, add $10^{(count-2)}$ to the *score* variable. This gives a score based on the number of repeated digits so far.
    d. Reset the *count* variable to 1 for the new digit.
    e. Set $prev\_digit$ to the current digit for the next iteration.
6. If the final digit(s) are repeated, add $2 \times 10^{(count-2)}$ to the *score* variable. This gives a bonus score for the final repeated digits.
7. Return *score*.

The Python code for assigning bonus to repeated positive integers is depicted in Code 4.7.

Code 4.7 Python code for assigning bonus to repeated positive integers

```
1  def duplicate_digit_score(n):
2      score = 0
3      # initialize previous digit to None
4      count = 1
5      prev_digit = None
6
7      '''
8      iterate over each digit
```

4.7 Bonus to Repeated Numbers

```
9        in n
10       '''
11       for digit in str(n):
12           '''
13           check if the current digit is
14           the same as the previous digit
15           '''
16           if prev_digit == digit:
17               # increment the count of repeated digits
18               count += 1
19           else:
20               '''
21               if the previous digit was also repeated,
22               score the count
23               '''
24               if count > 1:
25                   # score the count of repeated digits
26                   score += 10 ** (count - 2)
27                   # reset the count for the new digit
28                   count = 1
29           '''
30           update the previous digit
31           for the next iteration
32           '''
33           prev_digit = digit
34
35       '''
36       score any remaining repeated digits
37       at the end of the number
38       '''
39       if count > 1:
40           '''
41           double the score for the
42           last repeated digit
43           '''
44           score += 2 * 10 ** (count - 2)
45       return score
```

Table 4.8 The expected outputs for certain inputs to find nearest smaller numbers

Array	Expected output
[8,6,16,1921,17]	[6, 6, 6, 16, 16]
[2,3,4,871]	[2, 2, 3, 4]
[1,77,7,770,700,11]	[1, 1, 1, 7, 11, 7]
[6,8,9,888,1401,1402]	[6, 6, 8, 9, 888, 1401]

4.8 Nearest First Smaller Number

Let there be an array that is filled with positive integer numbers. In this challenge, for each number, it is aimed to find the first smallest number in the array, and if there is no smallest number, the number is to be unchanged. For example, consider $array = [2, 77, 13, 1]$. The first number is 2, and on the left of 2, there is no number, so the right part is considered. In the right, 1 is minimum, so $[2, 77, 13, 1] \rightarrow [1]$. The next number is 77. On the left of 2, there is 2, and on the right part 1 is minimum, but 2 is the first minimum, so $[2, 77, 13, 1] \rightarrow [1, 2]$. The next number is 13, and the first minimum is 1 $[2, 77, 13, 1] \rightarrow [1, 2, 1]$. The next number is 1, and no number is smaller than 1, the current element, which 1 is considered, so [1, 2, 1, 1]. Write a function that as input, takes a filled array with the integer numbers, and returns an array of numbers where each element in the array is the nearest smaller number to the corresponding element in the input array. For some inputs, the expected outputs are illustrated in Table 4.8.

Algorithm

The algorithm uses nested loops to iterate over each element in the input array and its surrounding elements. The algorithm finds the nearest smaller element for each element in the input array by comparing the element with its adjacent elements to the left and right of it. If either the left or right element is smaller than the current element, the smallest of the two elements is considered. It also handles the cases where the current element has no left or right element by setting the left or right element to the current element itself. The Python code for computing the nearest first smaller number is depicted in Code 4.8.

Code 4.8 Python code for computing the nearest first smaller number

```
1  def Nearest_first_smaller_number(array):
2      n = len(array)
3      nearest_smaller = []
4
5      # Iterate over each element in the input array
6      for x, current_element in enumerate(array):
7
8          '''
```

4.8 Nearest First Smaller Number

```
            Iterate over a range of indices to the
            left and right of the current element
            '''
            for y in range(1, n):

                '''
                Get the element to the left of the
                current element (or set it to the
                current element if there is no left element)
                '''
                if x >= y :
                    left = array[x - y]

                else:
                    left=current_element

                '''
                Get the element to the right of the
                current element (or set it to the
                current element if there is no right element)
                '''

                if x + y < n:
                    right = array[x + y]
                else:
                    current_element

                '''
                If either the left or right element
                is smaller than the current element,
                append the smallest of the two elements
                to the nearest_smaller list
                '''
                if left < current_element \
                or right < current_element:
                    nearest_smaller.append\
                    (left if left < right else right)
                    break

                '''
                If none of the elements to the left
                or right of the current element are
                smaller, append the current element
                itself to the nearest_smaller list
                '''
```

Table 4.9 The expected outputs for certain inputs to find an object that is preceded by k smaller numbers

Items,k	Expected output
[4,2, 1, 9, 14],k=3	9
[2,3,4,871], k=4	None
[700,3,900,400,1100],k=2	900
['cobol','ruby','c++','python','c','php'],k=2	'python'

```
54          else:
55              nearest_smaller.append(current_element)
56
57      return nearest_smaller
```

4.9 First Preceding by K Smaller Numbers

In this challenge, a list of objects is given and the task is to identify and return the object that is preceded by at least k number of smaller objects. In case such an item does not exist, the function will return a value of None. For example, consider $items = [2, 1, 9, 14]$, and $k = 2$. Which number is preceded by at least k smaller numbers? As 9 is preceded by 1 and 2, 9 is the answer. If the $items = [4, 2, 1, 9, 14]$ and $k = 2$, again the answer is 9, as the at least preceding is wanted. For the second example, consider ['$cobol'$, '$ruby'$, '$c++$', '$python'$, 'c', 'php'], and $k = 2$. Which string is preceded by at least k smaller numbers? Firstly, the length of each string is taken. The same operations as the previous example are done. The answer for the second one is '$python'$, as its length is six and preceded by '$c++$', '$cobol'$, and '$ruby'$. For some inputs, the expected outputs are illustrated in Table 4.9.

Algorithm

The algorithm goes through each item in the items list and checks if there are at least k smaller items before it. To do this, the algorithm creates a sublist of all items before the current item that are smaller than it. This is done using list comprehension. Once the sublist is created, its length is checked against k. If the length is greater than or equal to k, the algorithm returns the current item as the first item in the original list that has at least k smaller items before it. If no such item exists, the algorithm returns None. The Python code for computing first preceding number by smaller numbers is depicted in Code 4.9.

Code 4.9 Python code for computing first preceding number by smaller numbers

```
1   def first_preceded_by_k_smaller_number(items, k=1):
2       '''
3       Loop through each item in the list
4       'items' along with its index.
```

```
 5        '''
 6        for i, current_number in enumerate(items):
 7
 8            '''
 9            Create a list of all items before the current
10            item that are smaller than it.
11            '''
12            smaller_numbers = \
13            [n for n in items[:i] if n < current_number]
14
15            '''
16            If there are at least 'k' smaller items before
17            the current item, return the current item.
18            '''
19            if len(smaller_numbers) >= k:
20                return current_number
21
22        '''
23        If there is no item that satisfies
24        the condition, return None.
25        '''
26        return None
```

4.10 *n*-th Term of Calkin Wilf

The Calkin-Wilf tree is a rooted binary tree whose vertices are in one-to-one correspondence with the positive rational numbers. The root of the tree corresponds to the number 1, and for any rational number a/b, its left child corresponds to the number a/(a+b) and its right child corresponds to the number (a+b)/b. Each rational number appears exactly once in the tree. When performing a level-order traversal of the Calkin-Wilf tree, a sequence of rational numbers is generated, which is known as the Calkin-Wilf sequence. We would like to define a function that takes a positive integer n as input and returns the *n*-th term of the Calkin-Wilf sequence, which is a rational number. For example, consider the following example. If $n = 10$, the output is $\frac{3}{3+2}$. The process of reaching to the answer is described below. $\frac{1}{1}, \frac{1}{1+1}, \frac{1+1}{1}, \frac{1}{2+1}, \frac{2+1}{2}, \frac{2}{2+1}, \frac{2+1}{1}, \frac{1}{1+3}, \frac{1+3}{3}, \frac{3}{3+2}$. For some inputs, the expected outputs are illustrated in Table 4.10.

Table 4.10 The expected outputs for certain inputs to find *n*-th term in calkin-wilf

n	Expected output
11	5/2
2022	32/45
1993	59/43
24	2/7
10	3/5

Algorithm

It is used Breadth-First Search (BFS), which is a graph traversal algorithm to generate the Calkin-Wilf sequence up to the *n*-th term. The queue data structure is used to store the nodes of the binary tree representation of the sequence. The algorithm generates each term in the sequence by computing the left and right children of the current node and adding them to the queue. Finally, the *n*-th term of the sequence is returned. The following items illustrate the step-by-step process of the algorithm.

1. It takes a positive integer *n*.
2. Decrement *n* by 1.
3. Initialize an empty queue 'queue'.
4. Add the root node (1, 1) to the 'queue'.
5. Repeat the following steps *n* times:
    a. Dequeue the next node from the 'queue' using the dequeue function and assign its numerator and denominator to *parent_num* and *parent_denom*, respectively.
    b. Compute the left child of the current node by setting its numerator to *parent_num* and its denominator to *parent_num + parent_denom*.
    c. Compute the right child of the current node by setting its numerator to *parent_num + parent_denom* and its denominator to *parent_denom*.
    d. Add the left child and right child to the 'queue' using the enqueue function.
6. Dequeue the final node from the 'queue' using the dequeue function and assign its numerator and denominator to *final_num* and *final_denom*, respectively.
7. If the final node's denominator is 1, return the final node's numerator as an integer.

The Python code for computing and returning the *n*-th fractional term of calkin Wilf is depicted in Code 4.10.

Code 4.10 Python code for computing and returning the *n*-th fractional term of calkin wilf

```
1  def nth_term_calkin_wilf(n):
2      # A function to check if queue is empty
3      def is_empty(q):
4          return len(q) == 0
5
```

4.10 *n*-th Term of Calkin Wilf

```
6      # A function to add an item to the queue
7      def enqueue(q, item):
8          q.append(item)
9          return q
10
11     # Afunction to remove an item from the queue
12     def dequeue(q):
13         if not is_empty(q):
14             return q.pop(0)
15
16     '''
17     Initialize queue with the root node
18     (1/1) and decrement n by 1
19     '''
20     n -= 1
21     queue = []
22     queue = enqueue(queue, (1, 1))
23
24     '''
25     Generate the Calkin-Wilf sequence
26     up to the nth term using BFS
27     '''
28     for _ in range(n):
29         # Dequeue the next node from the queue
30         parent_num, parent_denom = dequeue(queue)
31         '''
32         Compute the left and right
33         children of the current node
34         '''
35         left_child_num = parent_num
36         left_child_denom=parent_num + parent_denom
37         right_child_num=parent_num + parent_denom
38         right_child_denom = parent_denom
39         '''
40         Enqueue the left and right
41         children of the current node
42         '''
43         queue = enqueue\
44         (queue, (left_child_num, left_child_denom))
45         queue = enqueue\
46         (queue, (right_child_num, right_child_denom))
47
48     # Dequeue the final node from the queue
49     final_num, final_denom = dequeue(queue)
50
```

```
51      # Return the nth term of the Calkin-Wilf sequence
52      if final_denom == 1:
53          return final_num
54      else:
55          return str(final_num) + '/' + str(final_denom)
```

4.11 Reverse Ascending Subarrays

In this challenge, an array of positive integer numbers is given, and the task is to find the reverse of the ascending subarrays. Each sub-list must be strictly ascending. For example, for [5, 7, 220, 33, 2, 6, 8, 1, 45] as input, output the reverse of the ascending subarrays: The subarrays are: [5, 7, 220], [33], [2, 6, 8], [1, 45], and in the next step, the sub-lists are reversed: [220, 7, 5,], [33], [8, 6,], [45, 1]. The reversed subarrays in their original order are merged. Hence, [5, 7, 220, 33, 2, 6, 8, 1, 45] → [220, 7, 5, 33, 8, 6, 2, 45, 1]. For some inputs, the expected outputs are illustrated in Table 4.11.

Algorithm

The algorithm use two-pointer technique, where 'i' pointer is used to point to the start of the sublist, and the 'j' pointer is used to find the end of the ascending sublist. After finding an ascending subarray, the program uses slicing and reversing to reverse the sublist elements and then appends them to the final result list. Finally, it sets the new starting index 'i' for the next subarray. The 'i' pointer is then incremented to the next sublist, and the process is repeated until the end of the list is reached. The Python code to reverse the ascending sub-lists is depicted in Code *reverse_ascending_subarrays*.

Code 4.11 Python code to reverse the ascending sub-lists

```
1   def reverse_ascending_subarrays(List_Numbers):
2       '''
3       create an empty list to store
4       the ascending sublists
5       '''
```

Table 4.11 The expected outputs for certain inputs to reverse the ascending subarrays

Arrays	Expected output
[6, 3, 2]	[6, 3, 2]
[5, 7, 220, 33, 2, 6, 8, 1, 45]	[220, 7, 5, 33, 8, 6, 2, 45, 1]
[1,5,7,9,90,13,11,23]	[90, 9, 7, 5, 1, 13, 23, 11]
[44,33,57,36,38,900]	[44, 57, 33, 900, 38, 36]

```
6       ascending_subarray = []
7       # set the starting index to 0
8       i = 0
9       # loop through the list
10      while i < len(List_Numbers):
11          # set j to the current index
12          j = i
13          # check if there is an ascending order
14          while j < len(List_Numbers) - 1 \
15          and List_Numbers[j] < List_Numbers[j+1]:
16              j += 1
17          '''
18          reverse and add the sublist
19          to the result list
20          '''
21          ascending_subarray.extend\
22          (List_Numbers[i:j+1][::-1])
23          # set the starting index for the next subarray
24          i = j + 1
25      return ascending_subarray
```

4.12 Smallest Integer Powers

Large integer powers can make calculations complicated in terms of both time and memory. The objective of this challenge is to find the smallest positive integers 'x' and 'y' such that a^x and b^y are within a certain tolerance of each other. In this context, 'tolerance' specifies the maximum acceptable difference between the two integer powers of 'a' and 'b'. Write a function that takes 'a', 'b', and 'tolerance' as input and returns the smallest positive integers 'x' and 'y'. For some inputs, the expected outputs are illustrated in Table 4.12.

Table 4.12 The expected outputs for certain inputs to find the smallest integer powers

a, b, tolerance	Expected output
12,16,10	(19, 17)
32,40,99	(248, 233)
43,33,73	(66, 71)

Algorithm

To solve this challenge, we utilize a greedy approach. The algorithm tries to find the smallest pair of positive integers x and y that satisfy the tolerance. It does this by iteratively increasing the values of x and y while also increasing the powers of a and b using repeated multiplication until the difference between the powers is less than or equal to a given tolerance. This approach may not always guarantee the optimal solution but it finds a reasonably good solution in a reasonably efficient way.

1. It takes integer number of a and b.
2. Set x and y to 1.
3. Set a_pow and b_pow to a and b, respectively.
4. Enter an infinite loop:

   a. If a_pow is greater than b_pow, check if the tolerance is less than or equal to b_pow divided by the difference between a_pow and b_pow.
   b. If it is, return x and y.
   c. Otherwise, increment y by 1 and multiply b_pow by b.
   d. If a_pow is less than b_pow, check if the tolerance is less than or equal to a_pow divided by the difference between b_pow and a_pow.
   e. If it is, return x and y.
   f. Otherwise, increment x by 1 and multiply a_pow by a. If a_pow and b_pow are equal, return x and y.

The Python code to compute and return the smallest integer powers is depicted in Code 4.12.

Code 4.12 Python code to compute and return the smallest powers

```
1  def smallest_integer_powers(a, b, tolerance=100):
2      x, y = 1, 1
3      a_pow, b_pow = a, b
4
5      while True:
6          if a_pow > b_pow:
7              if tolerance <= b_pow / (a_pow - b_pow):
8                  return x, y
9              y += 1
10             b_pow *= b
11         elif a_pow < b_pow:
12             if tolerance <= a_pow / (b_pow - a_pow):
13                 return x, y
14             x += 1
15             a_pow *= a
16         else:
17             return x, y
```

4.13 Sorting Cycles of a Graph

A cycle is a path that starts and ends at the same vertex. Consider a given array G, which is filled with positive integer numbers. Each number in G is the vertex of a graph. In this challenge, the task is to find the cycles G and sort them in some orders. For example, let $G = [2, 4, 6, 5, 3, 1, 0]$ be as input, find the cycles such that in each cycle, the greatest number be the first vertex in the cycle, and the vertices that their positions or indexes are after the greatest number position, are inserted after the greatest number position, while the vertices that their positions are before the greatest number position are inserted after the position of the greatest number and the vertices that their positions are after the greatest number position; after it, all cycles in ascending order must be sorted. The first vertex G is 2, which number is in index 2? As the index in Python starts from zero, so the number is 6. In the next step, which number is in the index 6? The number is 0. In the next step, which number is in index 0? The number is 0, so a cycle is found, 2→ 6→ 0 → 2. In this step, the first cycle is inserted, $C = [[2, 6, 0]]$. The first vertex that is not visited is 4, so it is considered the first vertex of the cycle. In the next step, which number is in index 4? The number is 3. In the next step, which number is in index 3? The number is 5. In the next step, which number is in index 5? The number is 1. In the next step, which number is in index 1? The number is 4. So, a cycle is found, 4→ 3→ 5 → 1 → 4. In this step, the second cycle is inserted, $C = [[2, 6, 0], [4, 3, 5, 1]]$. There is no remaining vertex in G, so the sorting process is to be performed. For [2, 6, 0], the greatest number (vertex) is 6, so [2, 6, 0] → [6], 0 is after 6, so [6] → [6, 0], and 2 is before 6, so [6, 0] → [6, 0, 2]. The updated cycles are $C = [[6, 0, 2], [4, 3, 5, 1]]$. For the second cycle [4, 3, 5, 1], the greatest number is 5, so [4, 3, 5, 1] → [5]. The next one after index 5 is 1, so 1 is inserted after 5 and previous ones after the index 5 are 4 and 3, so 4 and 3 are appended after 5 and 1, respectively. Hence the cycles are $C = [[6, 0, 2], [5, 1, 4, 3]]$. In the next step, the cycles in ascending order are sorted, so $C = [[6, 0, 2], [5, 1, 4, 3]] \to C = [[5, 1, 4, 3], [6, 0, 2]]$. In the last step, the nested array is flatted; $C = [5, 1, 4, 3, 6, 0, 2]$. Write a function that as input, takes array G, and returns the sorted cycles. For some inputs, the expected outputs are illustrated in Table 4.13.

Table 4.13 The expected outputs for certain inputs to sort cycles of graph

G	Expected output
[1,2,4,5,3,0]	[5, 0, 1, 2, 4, 3]
[1,2,3,4,0]	[4, 0, 1, 2, 3]
[4,3,7,0,1,5,2,6]	[4, 1, 3, 0, 5, 7, 6, 2]
[1,3,5,0,2,4]	[3, 0, 1, 5, 4, 2]

Algorithm

The algorithm is based on a combination of two techniques: depth-first search-based (DFS) and sorting. The DFS is used to detect all cycles in a directed graph. The algorithm 'findCycles' implements a DFS-based algorithm to detect cycles in the input graph. This algorithm visits each vertex in the graph only once and explores all outgoing edges from each vertex until it either finds a cycle. Once a cycle is found, it is added to the list of cycles and the algorithm continues with the next unvisited vertex. The algorithm 'sort' sorts each cycle in the input list of cycles according to a specific criterion. In this case, the criterion is to put the highest element of the cycle first and then append the remaining elements in order. Once all cycles are sorted, they are flattened into a single list. Finally, the sorted and flattened list of cycles is returned as the output. The Python code to find and sort the cycles of a given graph is depicted in Code 4.13.

Code 4.13 Python code to find and sort the cycles of a given graph

```
1   def insertion_sort(arr):
2       '''
3       Sort an array using the
4       insertion sort algorithm
5       '''
6       for i in range(1, len(arr)):
7           key = arr[i]
8           j = i - 1
9           while j >= 0 and key < arr[j]:
10              arr[j + 1] = arr[j]
11              j -= 1
12          arr[j + 1] = key
13      return arr
14  def Sorting_Cycles(graph):
15      # Find all cycles in a directed graph
16
17      flat_cycles = []
18      cycles = []
19
20      '''
21      DFS-based algorithm to detect
22      cycles
23      '''
24      def find_cycles(g):
25          visited = set()
26          for vertex in g:
27              if vertex in visited:
28                  continue
29              start = vertex
```

4.14 Obtaining Numbers in Balanced Ternary System

```
30                  current_cycle = [start]
31                  visited.add(start)
32                  neighbor = g[start]
33                  while neighbor != start:
34                      current_cycle.append(neighbor)
35                      visited.add(neighbor)
36                      neighbor = g[neighbor]
37                  cycles.append(current_cycle)
38          return cycles
39
40      '''
41      Sort a list of cycles
42      based on a specific criterion
43      '''
44      def sort_cycles(cycles):
45          sorted_cycles = []
46          for cycle in cycles:
47              if len(cycle) == 1:
48                  sorted_cycles.append(cycle)
49              else:
50                  highest = max(cycle)
51                  highest_index = cycle.index(highest)
52                  sorted_cycle = \
53                      cycle[highest_index:]+cycle[:highest_index]
54                  sorted_cycles.append(sorted_cycle)
55          sorted_cycles = insertion_sort(sorted_cycles)
56          flat_sorted_cycles = \
57              [vertex for cycle
58          in sorted_cycles for vertex in cycle]
59          return flat_sorted_cycles
60
61      # Call the find_cycles and sort_cycles functions
62      cycles = find_cycles(graph)
63      flat_cycles = sort_cycles(cycles)
64
65      # Return the flattened list of cycles
66      return flat_cycles
```

4.14 Obtaining Numbers in Balanced Ternary System

We know that in any base r, the digits are from 0 to $r - 1$. Let $n = 2201$, and convert it to base 10. $2 \times 3^0 + 2 \times 3^1 + 0 + 1 \times 3^3 = 35$. The calculations performed, were the normal way to convert from the base 3 to 10. There is another way to convert

Table 4.14 The expected outputs for certain inputs to obtain coefficients in the balanced ternary system

n	Expected output
25	[27, −3, 1]
28	[27, 1]
88	[81, 9, −3, 1]
1400	[2187, −729, −81, 27, −3, −1]

from the base 3 to 10. Consider the following example, $-1 \times 3^0 + 0 + 1 \times 3^2 + 1 \times 3^3 = 35$. The second approach so-called balanced ternary. Balanced ternary is a numeral system in those coefficients (digits) are consisted of 0, 1, and −1, while in the first approach, the coefficients consist of 0, 1, and 2. In this challenge, the task is to find the coefficients in the balanced ternary. Write a function that as input, takes a positive integer n, and returns the array of the coefficients in the balanced ternary, such that their multiplication by their position is equal n. Notice that the returned array must be in descending order, and for each element, the absolute number is considered. For some inputs, the expected outputs are illustrated in Table 4.14.

Algorithm

The algorithm takes a positive integer n as input. To obtain the coefficient, the algorithm computes the remainder of n divided by 3, denoted *Remainder*. It then considers the integer part of $(n + 1)/3$ and computes *Remainder* again. This process is repeated until n is not equal to 0. In the next step, let p be the corresponding position of each coefficient. For each element i in Remainder, the algorithm checks if i is equal to 1. If i is equal to 1, the algorithm computes 1 multiplied by 3 raised to the power of p. If i is equal to −1, the algorithm computes −1 multiplied by 3 raised to the power of p. If i is equal to 0, no computation is performed. The Python code for balanced ternary is depicted in Code 4.14.

Code 4.14 Python code for balanced ternary

```
1   def Obtaining_Numbers_in_Balanced_Ternary_System(n):
2       remainders = ""
3       while n!=0:
4           # Obtaining the remainder part
5           '''
6           The Coefficients are 0 ,1 and −1,
7           but −1 is two character one for − and another
8           for 1. Hence, 2 instead −1 is considered
9           '''
10          remainders = remainders+"012"[n % 3]
11          '''
12          [n % 3] generates a digit 0 or 1 or 2,
```

```
13            and returns the corresponding character "012".
14            i.e.
15            "012"[1]  ↔  ↔  ↔  ↔  ↔  1
16            "012"[0]  ↔  ↔  ↔  ↔  ↔  0
17            "012"[2]  ↔  ↔  ↔  ↔  ↔  2
18            '''
19            # Obtaining the integer part
20            n = ((n+1)//3)
21        Coefficients = remainders
22        Storage_Array = []
23        '''
24        Multiply each coefficient to
25        its corresponding position
26        '''
27        '''
28        0 is not considered, as multiplying it
29        with its position leads to zero
30        '''
31        for i in range(len(Coefficients)-1,-1,-1):
32            '''
33            The for loop in descending order
34            is considered because from the
35            largest position to the
36            smallest are considered.
37            '''
38            if Coefficients[i] == '1':
39                Storage_Array.append(3**i)
40                '''
41            2 is not recognized in the balanced ternary
42            if i is 2, so instead of 2 -1 is considered.
43            '''
44            elif Coefficients[i] == '2':
45                Storage_Array.append(-1*(3**i))
46        return Storage_Array
```

4.15 Is Strictly Ascending

The objective of this challenge is to check whether the elements of a given list are arranged in strictly ascending order. It is important to note that a list with duplicate numbers cannot be considered as strictly ascending. To implement the solution, please keep the following points in mind:

Table 4.15 The expected outputs for certain inputs to check if the a list of numbers is strictly ascending

Items	Expected output
[−6]	True
[1,2,4,7,1890]	True
[−3,−2,15,16,2000]	True
[−3,−2,159,18,915]	False

- The input to the problem will always be a list of numbers.
- A list with only one element, whether positive or negative, is considered to be in ascending order.
- In a strictly ascending list, there cannot be any duplicate numbers.

For a set of predefined inputs, the expected outputs are provided in Table 4.15.

Algorithm

The algorithm implemented in this problem involves each number in the input list checking itself against the previous number. If any number is found to be less than or equal to its preceding numbers, the algorithm will return false, indicating that the list is not strictly ascending. Conversely, if all numbers are strictly greater than their predecessors, the algorithm will return true, indicating that the list is indeed in strictly ascending order. The Python code for the strictly ascending problem is depicted in Code 4.15.

Code 4.15 Python code for the strictly ascending problem

```
1  def Is_Strictly_Ascending(items):
2      '''
3      Simply check each number to previous number
4      '''
5      '''
6      Considering the first number
7      as previous number
8      '''
9      previous=items[0]
10     templist=[]
11     for num in items:
12         # If the condition was met, return false
13         if num < previous:
14             return False
15         # If the condition was met, return false
16         if num in templist:
17             return False
18         #updating the previous number
19         previous=num
```

```
20          '''
21          If item is in templist,
22          there is duplicate
23          number and false is returned
24          '''
25          templist.append(num)
26      return True
```

4.16 Priority Sorting

In this challenge, you are given a list and a set, both of which contain positive and negative integers. Your task is to create a function that takes these two data structures as input and returns a sorted list that prioritizes the elements from the set. Specifically, the function should first identify the elements that are present in both the list and the set, and append them to a new list in ascending order. Then, it should append the remaining elements from the list to the same list, and sort the final list in ascending order. Write a function that as input, takes a list and a set, and returns a sorted list such the set element are to be in priority. For some inputs, the expected outputs are illustrated in Table 4.16.

Algorithm

The algorithm first identifies the minimum element in the input list and checks if it is present in the set. If it is, the algorithm appends the minimum element to the output list as many times as it appears in the input list, removes all occurrences of the minimum element from the input list, and removes the minimum element from the set to avoid duplicates. This process is repeated for all elements in the set, which ensures that the output list is sorted based on the priority set by the elements in the set. After all the elements in the set have been processed, any remaining elements in

Table 4.16 The expected outputs for certain inputs to sort with priority

List, set	Expected output
[5, 2,7, 3, 2, 1,29], {2, 7}	[2, 2, 7, 1, 3, 5, 29]
[1,4,8,9,13,11,17], {11, 12}	[11, 1, 4, 8, 9, 13, 17]
[4,5,7,13,290], {3, 4, 1, 8}	[4, 5, 7, 13, 290]
[4,−2,5,7,13,−8,9,8], {2, 4, 1, 8}	[−8, 4, −2, 5, 7, 8, 9, 13]

the input list are sorted and appended to the output list. The Python code for priority sort is depicted in Code 4.16.

Code 4.16 Python code for priority sort

```
def priority_sort(list, set):
    '''
    Sorts the input list in a specific
    way based on the elements in the set.
    If the list is empty, returns an empty list.
    '''

    # If the list is empty, return an empty list.
    if not list:
        return list

    # Create an empty list to store the sorted elements.
    sorted_list = []

    '''Create a range object that goes
    from 0 to the length of the set.
    '''
    iterate_count = range(len(set))

    # Loop through the range object.
    for i in iterate_count:
        '''Check if the minimum element in
        the set is also in the input list.
        '''
        if min(set) in list:
            '''
            If it is, loop through the input list
            and append the minimum element to sorted_list
            '''
            # as many times as it appears in input list.
            for k in range(list.count(min(set))):
                sorted_list.append(min(set))
                '''Remove all occurrences of the minimum
                element from the input list.
                '''
                list.remove(min(set))
            '''
            Remove the minimum element from
            the set to avoid duplicates.
```

4.17 Sorting Positives, Keep Negatives

```
40                  '''
41              set.remove(min(set))
42          else:
43              '''
44                  If the minimum element
45                  is not in the input list,
46                  simply remove it from the set.
47              '''
48              set.remove(min(set))
49
50      '''
51      Create a range object that goes from 0
52      to the length of the input list.
53      '''
54      iterate_count = range(len(list))
55
56      # Loop through the range object.
57      for i in iterate_count:
58          # Find the minimum element in the input list.
59          min_elem = min(list)
60          # Append the minimum element to sorted_list.
61          sorted_list.append(min_elem)
62          # Remove the minimum element from the input list.
63          list.remove(min_elem)
64
65      # Return the sorted list.
66      return sorted_list
```

4.17 Sorting Positives, Keep Negatives

In this challenge, the task is to sort the positive integer number and keep the order of negative integer numbers. Write a function that as input, takes an array of positive and negative numbers, and returns the array with the same length such the positive ones are sorted, while the order of negative ones is kept. For some inputs, the expected outputs are illustrated in Table 4.17.

Algorithm

Bubble sort is used to sort the positive numbers in the input array in ascending order, while keeping the negative numbers in their original positions. The dictionary is used to keep track of the indices and values of the negative numbers before they are

Table 4.17 The expected outputs for certain inputs to sort positive numbers while keep negatives

Array	Expected output
[79,7,−3,−4,6,−1]	[6, 7, −3, −4, 79, −1]
[451,419,−1001,2,3,−2]	[2, 3, −1001, 419, 451, −2]
[−39,−456,0,−7,−3,−599]	[−39,−456,0,−7,−3,−599]
[]	[]

removed from the array. After the positive numbers are sorted, the negative numbers are inserted back into their original positions using the indices stored in the dictionary. The Python code to sort positive numbers and keep negatives is depicted in Code 4.17.

Code 4.17 Python code to sort positives numbers and keep negatives

```python
def sort_positives_keep_negatives(array):
    # Sorting by bubble sort
    def bubblesort(array : list):
        for i in range(len(array)):
            flag = True
            # Iterate over array
            for j in range(len(array) - i - 1):
                '''
                Compare two adjcent numbers
                and swap if needed
                '''
                if array[j] > array[j + 1]:
                    array[j], array[j + 1] = \
                    array[j + 1], array[j]
                    flag = False
            if flag:
                return array
    # For keeping the negative items
    neg = {}
    for i in range(len(array)):
        # if current item is negative
        if array[i] < 0:
            '''
            add index:value into
            negatives dictionary
            '''
            neg[i] = array[i]
    # Temporarily remove negative items
    array = [i for i in array if i >= 0]
    # Sorting using bubblesort algorithm
    array=bubblesort(array)
```

```
32       # Iterate over dictionary
33       for i, v in neg.items():
34           '''
35           Insert negative numbers
36           in the corresponding index
37           '''
38           array.insert(i, v)
39       return array
```

4.18 Numbers First, Characters Second

In this challenge, while the structure of the given nested list is kept, the nested list must be sorted such that the numbers and then characters are to be sorted. Both numbers and characters must be sorted in ascending order. Write a function that as input, takes a nested list consisting of numbers and characters and returns a nested list with the same size of lists such that the numbers and then characters are sorted. For some inputs, the expected outputs are illustrated in Table 4.18.

Algorithm

The algorithm first flattens the nested list into a single list, then separates the numbers and characters, sorts them using Bubble Sort, and finally merges them back into a sorted list with the original structure of nested lists. The Python code to sort numbers and then characters is depicted in Code 4.18.

Code 4.18 Python code to sort positives numbers and keep negatives
```
1   def number_then_character(NestedList):
2       # Sorting by bubble sort
3       def bubblesort(lst):
4           for i in range(len(lst)):
5               flag = True
```

Table 4.18 The expected outputs for certain inputs to sort numbers first and characters second

NestedList	Expected output
[[0, 0, 0.5, 1], [3], [5, 5, 'X', 'Y'], ['Z', 'a'], ['b', 'er', 'f'], ['p', 's']]	[[0, 0, 0.5, 1], [3], [5, 5, 'X', 'Y'], ['Z', 'a'], ['b', 'er', 'f'], ['p', 's']]
[[2,74,3,0,0.5,],[4,7,'a','b','e',99]]	[[0, 0.5, 2, 3, 4], [7, 74, 99, 'a', 'b', 'e']]
[[419,419,'t','r','pol'],['x','y',88,'r']]	[[88, 419, 419, 'pol', 'r'], ['r', 't', 'x', 'y']]
[[2,3,'u'],['w',100,4,5],['r','d','t',1]]	[[1, 2, 3], [4, 5, 100, 'd'], ['r', 't', 'u', 'w']]

```python
         # iterate over list
         for j in range(len(lst) - i - 1):
             if lst[j] > lst[j + 1]:
                 '''
                 Compare two adjcent items
                 and swap if needed
                 '''
                 lst[j], lst[j + 1]= lst[j +1], lst[j]
                 flag = False
         if flag:
             # If no swap operation, so list is sorted
             return lst
    # Flatten the list
    flattedlist = [i for lst in NestedList for i in lst]
    # Store the size of each list
    sizes = []
    # Iterate over collection and save sizes
    for lst in NestedList:
        sizes.append(len(lst))
    # numbers : Extract numbers (int/float)
    numbers = \
    [n for n in flattedlist if isinstance(n, (int, float))]
    # characters: Extract characters
    characters = \
    [str(l) for l in flattedlist if str(l).isalpha()]
    # Sort seperated numbers
    numbers=bubblesort(numbers)
    # Sort seperated chars
    characters=bubblesort(characters)
    # Merge two sorted array
    flattedlist = numbers + characters
    # For storing final result
    sortedlist = []
    for s in sizes:
        # Create an array with size s
        sortedlist.append(flattedlist[:s])
        '''
        Update flattedlist, as the previous items
        are inserted into a list'''
        flattedlist = flattedlist[s:]
    return sortedlist
```

4.19 Sorting Dates

This challenge is aimed to sort the given date in the format $MM:HH_YYYY-MM-DD$ in an ascending or descending order. The priority criteria used for sorting are Year, month, day, hour, and minute. Write a function that as input takes the list of date times (as string) and the sort type and returns the sorted list of date times. For some inputs, the expected outputs are illustrated in Table 4.19.

Algorithm

It converts the given date times in format string to the date time object and sorts them using the bubble sort algorithm, and after sorting, the date times objects are converted to string object. The Python code to sort dates is depicted in Code 4.19.

Code 4.19 Python code to to sort dates

```
1   import datetime
2   def bubble_sort(arr):
3       for i in range(len(arr) - 1):
4           for j in range(0, (len(arr) - i) - 1):
5               if arr[j] > arr[j + 1]:
6                   # Swap if it is needed
7                   arr[j], arr[j + 1] = arr[j + 1], arr[j]
8       return arr
9   def sort_dates(times, sort_types):
10      time_objects = []
11      sorted_dates=[]
12      # Iterate through items
13      wanted_date='%d-%m-%Y_%H:%M'
14      for t in times:
15
16          time_objects.append(
```

Table 4.19 The expected outputs for certain inputs to sort the dates

times,sort_types	Expected output
['09-02-2001_10:03', '10-02-2000_18:29', '01-01-1999_00:55'],'ASC'	['01-01-1999_00:55', '10-02-2000_18:29', '09-02-2001_10:03']
['01-04-2004_10:03', '10-02-2006_03:29', '01-01-2022_00:55'],'ASC'	['01-04-2004_10:03', '10-02-2006_03:29', '01-01-2022_00:55']
['01-04-2004_10:03', '10-02-2006_03:29', '01-01-2022_00:55'],'DSC'	['01-01-2022_00:55', '10-02-2006_03:29', '01-04-2004_10:03']
['09-02-2001_10:03', '10-02-2000_18:29', '01-01-1999_00:55'],'DSC'	['09-02-2001_10:03', '10-02-2000_18:29', '01-01-1999_00:55']

```
17          datetime.datetime.strptime(t,wanted_date))
18      # Dates are sorted by bubble sort algorithm
19      time_objects=bubble_sort(time_objects)
20
21      if sort_types.lower() == 'asc':
22          for t in time_objects:
23              # Convert the data times to string
24              sorted_dates.append(
25                  (datetime.datetime.strftime(t,wanted_date)))
26          return sorted_dates
27      elif sort_types.lower() == 'dsc':
28          '''
29          As the data times are sorted in ascending order,
30          traversing from the last element is done.
31          '''
32          for t in time_objects[::-1]:
33              # Convert the data times to string
34              sorted_dates.append(
35                  (datetime.datetime.strftime(t,wanted_date)))
36          return sorted_dates
```

4.20 Sorting by Alphabetical Order and Length

This challenge is aimed to sort the given string and returns the sorted string such that at first sort them in alphabetical order and then sort them by length. Write a function that as input, takes a string and returns the sorted string in terms of alphabetical order and length, respectively. For some inputs, the expected outputs are illustrated in Table 4.20.

Table 4.20 The expected outputs for certain inputs to sort by alphabetical order and length

Sentence	Expected output
'Year of the Tiger, is it fog'	'Tiger, Year fog the it of is'
'Python is one of the most used languages'	'languages Python most used the one of is'
'FIFA World Cup Qatar'	'Qatar World FIFA Cup'
'for if while def range else set'	'range while else def set for if'

4.20 Sorting by Alphabetical Order and Length

Algorithm

The algorithm takes a sentence as input, splits it into words, sorts them in alphabetical order using the bubble sort algorithm, and then sorts them again based on their lengths. This is done by iterating over the sorted list of words and comparing each word's length with the lengths of the preceding words. If a word is longer than a preceding word, they are swapped. Finally, the sorted words are concatenated into a single string and returned.

The Python code to sort the given string by alphabetical order and length is depicted in Code 4.20.

Code 4.20 Python code to sort the given string by alphabetical order and length

```
1   def bubble_sort(arr):
2       for i in range(len(arr) - 1):
3           for j in range(0, (len(arr) - i) - 1):
4               if arr[j] > arr[j + 1]:
5                   # Swap if it is needed
6                   arr[j], arr[j + 1] = arr[j + 1], arr[j]
7       return arr
8   def sorted_by_alphabetical_and_length(sentence):
9       words = sentence.split()
10      # Sorting in alphabetical order
11      words=bubble_sort(words)
12      for word in range(0, len(words)):
13          for i in range(0, word):
14              # Swap if the condition was met
15              if len(words[word]) > len(words[i]):
16                  temp = words[i]
17                  words[i] = words[word]
18                  words[word] = temp
19
20      '''
21      Each sorted word is added to sorted_sentence
22      (convert the list to string)
23      '''
24      sorted_sentence = ""
25      for item in range(0, len(words)):
26          sorted_sentence += words[item] + ' '
27      return sorted_sentence
```

4.21 Sorting by Digit Count

In standard numerical comparison, 101 is greater than 9. However, in the context of sorting by digit count, 9 has a higher value than 101. The objective of this challenge is to sort the given numbers based on their digit count. For instance, consider an array $array = [81181972, 8111972]$, and let $a = 81181972$ and $b = 8111972$. The comparison starts from the most significant digit, which is 9 in this case. Both numbers have one 9, so the next significant digit, 8, is compared. There are two 8s in a and one in b, so a is considered as the greater number. The comparison continues in this manner: if there are equal digits in both a and b, the next significant digit is examined until the comparison reaches zero. Write a function that as input, takes an array of the positive integers, and returns the sorted array in terms of digit count. For some inputs, the expected outputs are illustrated in Table 4.21.

Algorithm

The 'divide-and-conquer' strategy is a problem-solving technique that involves breaking down a complex problem into smaller, more manageable sub-problems and solving each sub-problem individually. Merge sort algorithm is based on the 'divide-and-conquer' strategy. In the case of merge sort, the algorithm repeatedly divides a given list into multiple sublists until each sublist has only one item. These sublists are then merged together to form a sorted list. The algorithm first recursively divides the input array into two halves until the base case of a single element or an empty array is reached (merge sort). Then, it determines which of the two elements should come first in the sorted order based on the count of their digits. Finally, the two sorted sub-arrays are merged back together in sorted order. The Python code to sort the numbers in terms of counting the digits is depicted in Code 4.21.

Code 4.21 Python code to sort the numbers in terms of counting the digits

```
1  def sort_by_digit_count(arr):
2      """
3      Sorts an array of positive integers
4      by the count of their digits.
5      """
```

Table 4.21 The expected outputs for certain inputs to sort by digit count

Array	Expected output
[81181972,9,123198776,23,456,1]	[1, 23, 456, 9, 123198776, 81181972]
[1,2,3,11,9,77,66,87,111]	[1, 11, 111, 2, 3, 66, 77, 87, 9]
[39,456,0,7,3,599]	[0, 3, 456, 7, 39, 599]
[]	[]

4.21 Sorting by Digit Count

```
6        # Applying the algorithm of merge sort
7        if len(arr) > 1:
8            # Finding the mid of the array
9            mid = len(arr) // 2
10
11           # Dividing the array elements into 2 halves
12           L = arr[:mid]
13           R = arr[mid:]
14
15           # Sorting the first half
16           L = sort_by_digit_count(L)
17
18           # Sorting the second half
19           R = sort_by_digit_count(R)
20
21           # Merge the two sorted halves
22           i, j, k = 0, 0, 0
23           while i < len(L) and j < len(R):
24               if max_(L[i], R[j]) == R[j]:
25                   arr[k] = L[i]
26                   i += 1
27               else:
28                   arr[k] = R[j]
29                   j += 1
30               k += 1
31
32           '''Checking if any element was
33           left in the first half
34           '''
35           while i < len(L):
36               arr[k] = L[i]
37               i += 1
38               k += 1
39
40           '''Checking if any element was
41           left in the second half
42           '''
43           while j < len(R):
44               arr[k] = R[j]
45               j += 1
46               k += 1
47
48       return arr
49
50
```

```
51  def max_(a, b):
52      """
53      It takes two positive integers a and b,
54      and returns the one whose highest
55      digit occurs more times. In case of
56      a tie, returns the larger number.
57
58      """
59      a, b = str(a), str(b)
60      for i in range(9, −1, −1):
61          if a.count(str(i)) > b.count(str(i)):
62              return int(a)
63          elif a.count(str(i)) < b.count(str(i)):
64              return int(b)
65      return max(int(a), int(b))
```

Chapter 5
String

This chapter talks about 13 string-based problems. These challenges are explained with some examples and then programmed in Python. The problems are listed as follows:

1. Performing Pancake Scramble into texts
2. Reversing the vowels of a given text
3. Word shape from a Text Corpus
4. Word height from a Text Corpus
5. Combining adjacent colors based on given rules
6. McCulloch second machine
7. Champernowne Word
8. Combining the given strings into one new string based on some rules
9. Unscrambling the given words
10. Auto correcter word
11. Correct verb form in Spanish
12. Subsequent letters
13. Possible words from a text corpus.

5.1 Pancake Scramble into Texts

The pancake scramble problem is aimed to reverse characters of the given text t in the order 2, 3, ..., n, where n is the length of t. For example, t = '*python*', the steps to scramble it are: The first reversing from the second character is considered (y), '*python*' $\rightarrow$ '*ypthon*', another from third character (t) '*ypthon*' $\rightarrow$ '*tpyhon*' $\rightarrow$ '*hypton*' $\rightarrow$ '*otpyhn*', nth character '*otpyhn*' $\rightarrow$ '*nhypto*'. Write a function that as input, takes a string and applies pancake scramble to it. For some inputs, the expected outputs are illustrated in Table 5.1.

Table 5.1 The expected outputs for certain inputs for pancake scramble

t	Expected output
'python'	'nhypto'
'Deep Learning has revolutionized Pattern Recognition'	'.otnoe rta eiotlvrshgire eDepLann a eouinzdPtenRcgiin'
'Challenging Programming'	'gimroPginlaChlegn rgamn'
'x = 2x + 1(9 * 56y)'	')6 * (+2x = x195y'

Algorithm

It takes a string as input and returns a new string that is a scrambled version of the original string. It iterates over each index of the string and reverse the substring of the original string up to that index, and merge the reversed substring with the rest of the original string (i.e., the substring after the index) to create a new string. The algorithm repeats this process for each index of the string and returns the final scrambled string. The Python code to apply the pancake scramble in the given text is depicted in Code 5.1.

Code 5.1 Python code to apply the pancake scramble in the given text

```
1   # a function that reverses a string using a for loop
2   def reverse_string(s):
3       reversed_s = ''
4       # Iterate over indices of s backwards
5       for i in range(len(s)-1, -1, -1):
6           '''
7           Append character at the
8           current index to reversed_s
9           '''
10          reversed_s += s[i]
11      # Return fully reversed string
12      return reversed_s
13
14  # Define the pancake_scramble function
15  def pancake_scramble_into_texts(t):
16      for i in range(len(t)):
17          # Reverse substring up to index i
18          Reversed_String = reverse_string(t[:i+1])
19          # Slice the rest of the string
20          Orginal_String = t[i+1:]
21          # Merge the reversed and original substrings
22          t = Reversed_String + Orginal_String
23      # Return the fully scrambled string
24      return t
```

5.2 Reverse Vowels into Texts

Table 5.2 The expected outputs for certain inputs for reversing vowels

t	Expected output
'Other'	'Ethor'
'Deep Learning has revolutionized Pattern Recognition'	'Doip Liorneng hes ravelitoinuzod Pettarn Ricagneteen'
'Challenging Programming'	'Chillangong Prigremmang'
'x = 2x + 1(9 * 56y)'	')6 * (+2x = x195y'

5.2 Reverse Vowels into Texts

In this challenge, a text is provided and a new text is generated in which the vowels are reversed. Additionally, if a character at a specific index in the original text is uppercase, the corresponding character in the new string will also be uppercase. For example, if $s = Other$ is the given text, $o = Ethor$ is the new string generated that its vowels are reversed. In this problem, the considered vowels are 'aeiouAEIOU'. For some inputs, the expected outputs are illustrated in Table 5.2.

Algorithm

For reversing vowels in a string, the algorithm performs iteration over the characters of the input string, with additional iterations over the indices of the relevant vowel characters. Specifically, the algorithm first extracts all the vowel characters and their indices from the input string using a list comprehension that filters the characters by membership in a set of vowels. Then, it iterates over the indices in reverse order and collects the reversed vowels into a list. Finally, the program iterates over the original vowel positions within the input string and replaces them with their corresponding reversed vowels, while preserving the uppercase format of each vowel character within its respective position. The Python code for reversing the vowels of the given text is depicted in Code 5.2.

Code 5.2 Python code for reversing the vowels of the given text

```python
def reverse_vowels_into_texts(s):
    # Set of vowels
    vowels = set("aeiouAEIOU")
    chars = list(s)
    vowel_indices = \
        [i for i in range(len(chars)) if chars[i] in vowels]

    # extract all the vowels and their indices
    reversed_vowels = []
    for i in range(len(vowel_indices)-1, -1, -1):
        reversed_vowels.append(chars[vowel_indices[i]])

```

Table 5.3 The expected outputs for certain inputs for word shape

Words, shape	Expected output
['congeed', 'outfeed', 'strolld', 'mail', 'stopped'], [1, −1, −1, −1, 0, −1]	['congeed', 'outfeed', 'strolld']
['eeten', 'good', 'aare', 'oozes', 'sstor'], [0, 1, −1, 1]	['eeten', 'oozes', 'sstor']

```
13          ' ' '
14          Replace the original vowel
15          positions with the reversed vowels
16          ' ' '
17          j=0
18          for i in vowel_indices:
19              if chars[i].isupper():
20                  chars[i] = reversed_vowels[j].upper()
21              else:
22                  chars[i] = reversed_vowels[j].lower()
23              j+=1
24
25          return ''.join(chars)
```

5.3 Word Shape from a Text Corpus

The shape of a given word w in length m is a list l that consisted of $m-1$ of $+1, -1$, and 0. The way that an integer is to be appended in l, depends on the alphabetical order. The order is 'a' < 'b' < 'c' < 'd' < 'e' < 'f' < 'g' < 'h' < 'i' < 'j' < 'k' < 'l' < 'm' < 'n' < 'o' < 'p' < 'q' < 's' < 't' < 'u' < 'v' < 'w' < 'x' < 'y' < 'z'. For each alphabetical pair α and β, if $\alpha < \beta$ $shape = 1$, if $\alpha > \beta$ $shape = -1$, and if $\alpha = \beta$, $shape = 0$. For example, if the given $word = optimum$, what is the shape? The steps to reach the shape are: $o < p$, so $shape = [1] \to p < t$, so $shape = [1, 1] \to t > i$, so $shape = [1, 1, -1] \to i < m$, so $shape = [1, 1, -1, 1] \to m < u$, so $shape = [1, 1, -1, 1, 1] \to u > m$, so $shape = [1, 1, -1, 1, 1, -1]$. In this challenge, all words whose shapes are equal to the given shape must be found. Write a function that as input, takes a list of words (text Corpus) and shape, and return the list of words whose shape is exactly the same as the given shape. For some inputs, the expected outputs are illustrated in Table 5.3.

Algorithm

The algorithm applies a set of rules to determine the shape of a word, based on the alphabetical order of its characters, and uses this shape to filter a list of words. The rules are implemented using if-else statements. The Python code for computing and returning all equal shapes with the given shape is depicted in Code 5.3.

5.4 Word Height from a Text Corpus

Code 5.3 Python code for computing and returning the all equal shapes with the given shape

```
1  def Word_Shape_from_a_Text_Corpus(words, shape):
2      output = []
3      # To iterate through the list of words
4      for i in words:
5          '''
6          Just consider the words that
7          equals with given shape
8          '''
9          if len(i) == len(shape)+1:
10             if word_shape(i) == shape:
11                 output.append(i)
12     return output
13 def word_shape(word):
14     # To create an array
15     shape = ([None] * (len(word) - 1))
16     for i in range((len(shape))):
17         '''
18         Ord as inpute takes a character and
19         returns an integer (unicode) such that
20         the alphabetical order is considered.
21         '''
22         '''
23         Determine the shape for a given word
24         using the ruled based criteria
25         '''
26         a = ord(word[i])
27         b = ord(word[i+1])
28         if (a<b):
29             shape[i] = 1
30         elif (a==b):
31             shape[i] = 0
32         elif (a>b):
33             shape[i] = -1
34     return shape
```

5.4 Word Height from a Text Corpus

The length of a word refers to the number of characters it contains, while the height of a word indicates the number of meaningful sub-words that can be derived from it. A word with no inherent meaning has a height of zero, whereas a word that is meaningful and cannot be split into two meaningful sub-words has a height of one.

Table 5.4 The expected outputs for certain inputs for word height

Words, word	Expected output
'A text corpus', wjobnv	0
'A text corpus', chukker	1

For words that can be broken down into sub-words, the height of the word is equal to the highest height of its sub-words plus one. For example, the word 'roqm' has no inherent meaning and, therefore, a height of zero. On the other hand, the word 'chukker' cannot be split into two meaningful sub-words, so its height is one. Finally, for a word like 'enterprise' we can recursively break it down into sub-words until we reach meaningless words, and its height is determined by the maximum height of its sub-words. Write a function that as input, takes a list of words, denoted by $words$ and a single word, denoted by $word$ and returns the height of the word, where $word$ for finding its sub-words searches into $words$. For some inputs, the expected outputs are illustrated in Table 5.4.

It is important to note that the height of a word is determined based on the words in the text corpus.

Algorithm

This algorithm recursively computes the height of a word by splitting it into smaller parts, computing the heights of those parts recursively, and combining the heights of the parts to obtain the height of the original word. Memoization is used to avoid redundant computations and improve performance. The algorithm steps are outlined in detail as follows:

1. It takes three inputs: 'words', a list of words in the text corpus, 'word', the word whose height needs to be computed, and 'memo', an optional dictionary used to memorize previously computed heights.
2. If 'memo' is not provided, an empty dictionary is created for memoization.
3. If the height of the current word has already been computed and stored in the 'memo', the memoized value is returned.
4. A binary search algorithm is utilized to search for a given word in the list of words. If the word is found, the index of the word in the list is returned; otherwise, −1 is returned.
5. If the word is not present in the list of words, its height is 0, and 0 is returned, and the algorithm is terminated.
6. A list, 'validList', is created to store all possible ways to split the word into two parts and check if both parts are present in the list of words.
7. A loop iterates over all possible split positions in the word, from position 1 to the length of the word −1. For each split position, the left and right parts of the word are extracted.
8. If both the left and right parts are found in the list of 'words', they are added to the 'validList'.
9. If no valid splits are found, the height of the word is 1, and 1 is returned. 1

5.4 Word Height from a Text Corpus

10. A list, 'hs', is created to store the heights of all valid splits.
11. For each valid split (ls, rs), the height of the left and right parts is computed recursively. If the height of a part has already been computed and stored in the 'memo', the memoized value is used; otherwise, it is called recursively to compute the height. The heights of both parts are added together, and 1 (the original word is meaningful) is added to obtain the height of the current split.
12. The maximum height of all valid splits is stored in the 'memo' for the current word, and it is returned.

The Python code to compute the height of a word is depicted in Code 5.4.

Code 5.4 Python code to compute the word height

```
'''
This function implements binary search
to search for a value in a sorted list
'''
'''
Binary search is a popular algorithm used for
searching an ordered list of elements.
'''
'''
It works by repeatedly dividing the search interval
 in half until the target value is found or
the search interval becomes empty.
'''
def binary_search(lst, x):
    left = 0
    right = len(lst) - 1

    while left <= right:
        mid = (left + right) // 2
        if lst[mid] < x:
            left = mid + 1
        else:
            right = mid - 1

    return left


'''
This function computes the height
of a given word in a list of words
by recursively computing the height
of its subwords
'''
def Word_Height_From_a_Text_Corpus(words, word, memo=None):

```

```
35        # Create an empty memo if it doesn't exist
36        if memo is None:
37            memo = {}
38
39        '''
40        Check if the height of the current word has
41        already been computed and stored in the memo
42        '''
43        if word in memo:
44            return memo[word]
45
46        '''
47        searching for a word in the list of
48        words using binary search
49        '''
50        def Search(l, x):
51            i = binary_search(l, x)
52            return i if i != len(l) and l[i] == x else -1
53
54        '''
55        If the word is not present in the list
56        of words, its height is 0
57        '''
58        if Search(words, word) == -1:
59            return 0
60
61        validList = []
62        '''
63        Find all possible ways to split the
64        word into two parts and check if both
65        parts are present in the list of words
66        '''
67        for s in range(1, len(word)):
68            ls = word[0:s]
69            rs = word[s:]
70
71            if Search(words, ls) != -1 \
72            and Search(words, rs) != -1:
73                validList.append((ls, rs))
74
75        '''
76        If no valid splits are found,
77        the height of the word is 1
78        '''
79        if not validList:
```

5.5 Color Combination

```
80              return 1
81
82          hs = []
83          '''
84          Compute the height of each valid split
85          recursively and take the maximum
86          '''
87          for (ls, rs) in validList:
88            left = memo.get(ls, None) or \
89              Word_Height_From_a_Text_Corpus(words, ls, memo)
90            right = memo.get(rs, None) or \
91              Word_Height_From_a_Text_Corpus(words, rs, memo)
92            hs.append(max(left, right) + 1)
93
94          '''
95          Store the computed height of the current word
96          in the memo before returning it
97          '''
98          memo[word] = max(hs)
99          return max(hs)
```

5.5 Color Combination

In this challenge, three colors *yellow*, *red*, and *blue* are given, whose rules are as follows: (1) if two similar colors are added, the output is equal to the similar color (2) if two different colors are added, the output is the third color. For example, if the given string is $s = $ '*rybyr*', so determines the output. From the most left index, pair 0 and 0 + 1 are considered, *r* and *y* leads to reach *b* and *b* is inserted in array *A*, the next pair is 1 and 1 + 1, *y* and *b* leads to reach *r*, and *r* is inserted in array *A*, the next pair is 2 and 2 + 1, *b* and *y* leads to reach *r*, and *r* is inserted in array *A*, the next pair is 3 and 3 + 1, *y* and *r* leads to reach *b*, and *b* is inserted in array *A*, $A = brrb$. From the most left index, pair 0 and 0 + 1 are considered, *b* and *r* leads to reach *y* and *y* is inserted in array *B*, the next pair is 1 and 1 + 1, *r* and *r* leads to reach *r* and *r* is inserted in array *B*, the next pair is 2 and 2 + 1, *r* and *b* leads to reach *y*, and *y* is inserted in array *B*; $B = yry$. From the most left index, pair 0 and 0 + 1 are considered, *y* and *r* leads to reach *b* and *b* is inserted in array *C*, the next pair is 1 and 1 + 1, *y* and *r* leads to reach *b*, *b* is inserted in array *C*, $C = bb$. From the most left index, pair 0 and 0 + 1 are considered, *b* and *b* leads to reach *b* and *b* is inserted in the array *D*. The length *D* is one and, the only value in *D* is to be returned, so *b* is the answer. Write a function that as input, takes a string consisting of colors and returns one color from the combinations of the colors. For some inputs, the expected outputs are illustrated in Table 5.5.

Table 5.5 The expected outputs for certain inputs to combine adjacent colors

Colors	Expected output
'rybyr'	'b'
'rrrryybby'	'y'
'rby'	'b'
'bbbryryrybrrbyr'	'y'

Algorithm

It takes a list of *colors* as input and returns a single color. It does this by repeatedly combining adjacent pairs of colors, until only one color remains. Specifically, it divides the input list of colors into smaller lists of colors, and then combines each pair independently. That final color is then returned as the output of the function. To avoid recomputing the same combinations over and over again, the function stores the results of previous combinations in a dictionary so that it can use them again if needed. The Python code to find the last color of combined adjacent colors is depicted in Code 5.5.

Code 5.5 Python code to find the last color of combined colors

```
1   def combine(a,b):
2       # Make a dictionary to find the value that is to be returned.
3       ColorFinder = {"b":{"y":"r", "r":"y"},
4                      "y":{"b":"r", "r":"b"},
5                      "r":{"b":"y", "y":"b"}}
6       '''If the colors are the same,
7       so  return one of them. otherwise,
8       use the dictionary to find the proper clor.
9       '''
10      if (a==b):
11          return a
12      else:
13          return ColorFinder[a][b]
14  '''
15  This function takes a list of colors as input,
16  combines adjacent colors according to a set of
17  rules, and returns a single color.
18  '''
19  def combine_adjacent_colors(colors):
20      '''
21      Create a dictionary to store results of
22      previous calls to combine().
23      The keys are tuples of color pairs,
24      and the values are the combined colors.
```

5.5 Color Combination

```
25      '''
26      cache = {}
27
28      '''
29      Keep looping until there is only one
30      color left in the list.
31      '''
32      while len(colors) > 1:
33          '''
34          Create a new list to store the results
35          of combining adjacent colors.
36          '''
37          TemparrayColors = ([None] * (len(colors) - 1))
38
39          '''
40          Loop over adjacent pairs of colors
41          in the input list.
42          '''
43          for i in range(len(TemparrayColors)):
44              '''
45              Check if the result for the current
46              color pair is already in the cache.
47              '''
48              if (colors[i], colors[i+1]) in cache:
49                  # If it is, use the cached result.
50                  TemparrayColors[i] = \
51                      cache[(colors[i], colors[i+1])]
52              else:
53                  '''
54                  If it is not, call the combine() function
55                  to get the result, store it in the cache,
56                  and use it.
57                  '''
58                  result = combine(colors[i], colors[i+1])
59                  cache[(colors[i], colors[i+1])] = result
60                  TemparrayColors[i] = result
61
62          '''
63          Replace the input list with the new
64          list of combined colors.
65          '''
66          colors = TemparrayColors
67      # Return the final color.
68      return colors[0]
```

Table 5.6 The expected outputs for certain inputs to rewrite the digits

X	Expected output
'2999'	'999'
'329'	'929'
'101'	None
'322097845'	'209784522097845'

5.6 McCulloch Second Machine

In this challenge, an example of esoteric programming languages is being considered to boost your skills in Python programming. An esoteric programming language, also known as an esolang, is a programming language that is designed to test complex and unconventional ideas. These languages are not intended for practical use but are used for entertainment or amusement purposes. The Mcculloch string rewriting system utilizes the digits 0 to 9, with specific rules being applied to these digits. In this context, the X array represents strings with values ranging from 0 to n − 1, whereas the Y array is computed from X and consists of values ranging from 1 to n − 1.

1. If $X[0] = 2$, Y is the rest of the array X that is to be returned.
2. If $X[0] = 3$, $Y + 2Y$.
3. If $X[0] = 4$, $Y[:: -1]$.
4. If $X[0] = 5$, $Y + Y$.

Write a function that as input, takes string X, and returns the rewritten string based on the mentioned rules. For some inputs, the expected outputs are illustrated in Table 5.6.

Algorithm

An algorithm is used to rewrite a given string according to certain rules. This algorithm takes an input string X and, based on its first digit, selects a corresponding rule to apply to the digits. This process is then repeated recursively until there are no more digits left in the string or until an digit that is not covered by the rules is encountered. The Python code to rewrite the given string is depicted in Code 5.6.

Code 5.6 Python code to rewrite the given string

```
1  # A function to remove the first digit
2  def remove_first_digit(X):
3      return X[1:]
4
5  '''
6  A function to slice string and
7  insert '2' in the middle
8  '''
```

```
 9  def slice_and_append_2(X):
10      return mcculloch(X[1:]) + '2' + mcculloch(X[1:])
11
12  # A function to reverse the string
13  def reverse(X):
14      return mcculloch(X[1:])[::-1]
15
16  # A function to double the slices
17  def double(X):
18      return mcculloch(X[1:]) + mcculloch(X[1:])
19
20  def mcculloch(X):
21      '''
22      A dictionary that maps the first
23      digit to the corresponding function
24      '''
25      functions = {
26          '2': remove_first_digit,
27          '3': slice_and_append_2,
28          '4': reverse,
29          '5': double,
30      }
31      # If the first digit of X is in the dictionary
32      if X[0] in functions:
33          '''
34          Call the corresponding function with
35          the remaining digits of X as input
36          '''
37          return functions[X[0]](X)
```

5.7 Champernowne Word

The Champernowne word is a lengthy string that does not contain comma separators. It is composed of a sequence of numbers that begins with one and increases by one. The objective of this challenge is to return the corresponding number for a given index. It is important to note that this problem cannot be solved using the conventional method, such as *list.index*, due to time and space constraints. To address this challenge, a function should be written that receives a positive integer n as input, representing the desired index or position, and returns the corresponding number. For instance, in the seventh digit of the sequence '1, 2, 3, 4, 5, 6, 7, 8, 9, 10, ...', there is the digit 8. The expected outputs for some inputs are presented in Table 5.7.

Table 5.7 The expected outputs for certain inputs for champernowne word

X	Expected output
12**214	'9'
7	'8'
71789798769185258877 0047	'2'
3111111198765431001	'1'

To solve this challenge, the number of digits in each sequence is considered, as illustrated in the below table.

Sequence	Range	Number of words	Number of digits in each sequences
1	1–9	9	9
2	10–99	90	180
3	100–999	900	2700
4	1000–9999	9000	36,000

Algorithm

1. It takes one input parameters: *n* (the *n*th digit we want).
2. Initialize the variables 'd' (digits), 'p_i' (passed indexes), 'p_n' (passed numbers), and 'w' (number of words) to 0 and 9, respectively.
3. While 'n' is greater than or equal to 'p_i + w * (d + 1)', increment 'd' by 1, update 'p_i' to 'p_i + w * (d + 1)', update 'p_n' to 'p_n + w', and update 'w' to 'w * 10'.
4. Calculate the index 'p' as '(n − p_i) // (d + 1)'.
5. Calculate the wanted number as 'p_n + p + 1'.
6. Convert the wanted number to a string and find the character at index '(n − p_i) % (d + 1)'.
7. Return nth digit as the result of the algorithm.

The Python code for Champernowne word is depicted in Code 5.7.

Code 5.7 Python code for Champernowne word

```
1  def Champernowne_Word(n):
2      d = 0
3      p_i = 0
4      p_n = 0
5      w = 9
6
7      while n >= p_i + w * (d + 1):
8          p_i += w * (d + 1)
```

5.8 Combining the Strings into One New String

```
9          p_n += w
10         d += 1
11         # Considering another sequence
12         w *= 10
13
14     p = (n - p_i) // (d + 1)
15
16     '''
17     Convert it to string and obtain the
18     index of the digit we want
19     '''
20
21     num = str(p_n + p + 1)
22     return num[(n - p_i) % (d + 1)]
```

5.8 Combining the Strings into One New String

Let $v = \{aeiou\}$ be the set of vowels, and X and Y be the first and second words, respectively. X has one group of vowels if at least $i \in v, i$ exist in x, but if $i + 1, i + 2, i + 3, i + 4$ exist in x, there is one group of vowels, too. For example, *pyth**o**n**i**st* has two groups of vowels, while word *k**ee**p* has one group of vowels. This challenge is aimed at the two given strings generating a new one. The rules for generating the new string are:

1. If there is one vowel group in X just the characters before the vowel are kept, and it concated with Y such that the first consonants in Y before the first vowel are removed. For example, if $X = $ '*go*' and $Y = $ '*meaning*' be the given string strings, the output is '*geaning*'.
2. If there is more than one vowel group in X just the characters before the second vowel are kept, and it concated with Y such that the first consonants in Y before the first vowel are removed. For example, if $X = $ '*python*' and $Y = $ '*visual*' be the given string strings, the output is '*pythisual*'.

Write a function that as input, takes two strings X and Y, and returns a new string by a combination of X and Y with considering the rules above. For some inputs, the expected outputs are illustrated in Table 5.8.

Algorithm

It takes two input strings, *first_word* and *second_word*, and combines them into a single string. The combination is done by selecting appropriate segments of each input string. The first step in the algorithm is to identify the consecutive vowels in each input string. This is done by iterating over each character in the string and

Table 5.8 The expected outputs for certain inputs for combining the strings

First, second	Expected output
'go', 'meaning'	'geaning'
'python', 'visual'	'pythisual'
'elliot', 'bill'	'ill'
'ross', 'jules'	'rules'

checking if it is a vowel. If a character is a vowel, the algorithm checks if the previous character was also a vowel. If it was, the index of the current vowel is added to the previous sublist of consecutive vowels. Otherwise, a new sublist is created for the current vowel. Once the consecutive vowel groups have been identified for each input string, the algorithm determines which segments of the input strings to keep in the combined string. If the first input string contains no consecutive vowels, the entire string is kept. If the first input string contains only one group of consecutive vowels, the algorithm keeps everything in the first input string up to the first vowel. Otherwise, the algorithm keeps everything in the first input string up to the second-to-last group of consecutive vowels. The segment of the second input string to keep in the combined string always starts at the first vowel of the second input string. Finally, the algorithm combines the selected segments of the two input strings and returns the resulting string. The Python code for combining the strings into one new string is depicted in Code 5.8.

Code 5.8 Python code for combining the strings into one new string

```
1   def combine_strings_into_one(first_word, second_word):
2       vowels = 'aeiou'
3
4       '''
5       Extract consecutive vowels from the first word
6       '''
7       first_vowel_groups = []
8       for i in range(len(first_word)):
9           if first_word[i] in vowels:
10              '''
11              If the previous character is also a vowel,
12              add the index to the previous sublist
13              '''
14              if i > 0 and first_word[i-1] in vowels:
15                  first_vowel_groups[-1].append(i)
16              '''
17              Otherwise, create a new sublist
18              for the current vowel
19              '''
20          else:
```

5.8 Combining the Strings into One New String

```
21                    first_vowel_groups.append([i])
22
23      '''
24      Extract consecutive vowels
25      from the second word
26      '''
27      second_vowel_groups = []
28      for i in range(len(second_word)):
29          if second_word[i] in vowels:
30              '''
31              If the previous character is also
32              a vowel, add the index to the
33              previous sublist
34              '''
35              if i > 0 and second_word[i-1] in vowels:
36                  second_vowel_groups[-1].append(i)
37              '''
38              Otherwise, create a new sublist
39              for the current vowel
40              '''
41              else:
42                  second_vowel_groups.append([i])
43
44      # Determine the segment of the first word to keep
45      '''
46      If there are no consecutive vowels
47      in the first word, keep the whole word
48      '''
49      if len(first_vowel_groups) == 0:
50          fw_segment = first_word
51      '''
52      If there is only one group of
53      consecutive vowels, keep everything
54      in the first word up to the first vowel
55      '''
56      elif len(first_vowel_groups) == 1:
57          fw_segment = first_word[:first_vowel_groups[0][0]]
58      '''
59      Otherwise, keep everything in the
60      first word up to the second-to-last
61      group of consecutive vowels
62      '''
63      else:
64          fw_segment = first_word[:first_vowel_groups[-2][0]]
65
```

```
66            '''
67            Determine the segment of the second
68            word to keep, which starts at the
69            first vowel of the second word
70            '''
71            sw_segment = second_word[second_vowel_groups[0][0]:]
72
73        return fw_segment + sw_segment
```

5.9 Unscrambling the Given Word

English words can be scrambled arbitrarily without affecting their readability, as long as the first and last letters remain the same and the length of the word remains unchanged. The goal of this challenge is to unscramble a given word and identify similar words from a provided list. Write a function that takes a string 'word' and a list of words 'words' as inputs and returns a list of words from 'words' that meet the following criteria: (1) The word has the same length as 'word' and the first and last letters are the same as those of 'word'. (2) The sorted order is the same. For some inputs, the expected outputs are illustrated in Table 5.9.

Algorithm

Suppose we are given a string 'word' and a list of words 'words'. To unscramble 'word', we perform the following steps:

1. Create an empty array 'res' to store the similar words.
2. Iterate over each word 'w' in 'words' using a linear search.
3. For each word 'w', check if it has the same length as 'word' and if its first and last letters are the same as those of 'word'.
4. If the checks in previous steps are successful, check if the first and last letters are the same as the letters of 'word'.
5. If the checks in steps 3 and 4 are successful, add 'w' to the 'res' array.
6. Repeat steps 3–5 for all words in 'words'.
7. Return the 'res' array containing the similar words.

Table 5.9 The expected outputs for certain inputs to unscramble a given word

Words, word	Expected output
['pycorn', 'pipline', 'python', 'ceo', 'we'], 'pohytn'	['python']
['camerier', 'academic', 'company', 'creamier'], 'ceamierr'	['camerier', 'creamier']

5.10 Auto-Correcter Word

The Python code to unscramble a given word is depicted in Code 5.9.

Code 5.9 Python code to unscramble a given word

```
1  def bubble_sort(arr):
2      for i in range(len(arr) - 1):
3          for j in range(0, (len(arr) - i) - 1):
4              if arr[j] > arr[j + 1]:
5                  # Swap if it is needed
6                  arr[j], arr[j + 1] = arr[j + 1], arr[j]
7      return arr
8  def Unscrambling the given words(words, word):
9      res = []
10     for w in words:
11         '''
12         By using the continue statement, the
13         current w is skipped and the next one is considered.
14         '''
15         if len(w) != len(word):
16             continue
17         elif w[0] != word[0] or w[-1] != word[-1]:
18             continue
19         # You can employ any sorting algorithm
20         if bubble_sort(list(word)) == bubble_sort(list(w)):
21             res.append(w)
22     return res
```

5.10 Auto-Correcter Word

To auto-correct words, several methods can be used. One of the most widely used techniques is keyword distance. This method involves replacing a word with the closest letter to the intended word based on their proximity on a keyboard. For instance, if the intended word is 'car', but the input word is 'csr', the letter 'a' which is adjacent to 's' on the keyboard should replace 's', so 'csr' → 'car'. This challenge requires computing all keyword distances between the given word and other words in a list, and then replacing the given word with the word that has the minimum keyword distance. Write a function that as input, takes a string $word$, a list of words $words$, and for each w in $word$, returns w that its distance against $word$ is minimum. For some inputs, the expected outputs are illustrated in Table 5.10.

Algorithm

Suppose we are given a string 'word' and a list of words, denoted by 'words'. For each word 'w' in words, the algorithm checks if 'w' and 'word' have the same length.

Table 5.10 The expected outputs for certain inputs to auto-correct a given word

Words, word	Expected output
['pycorn', 'pipline', 'python', 'ceo', 'we'], 'pohytn'	'python'
['camerier', 'academic', 'company', 'creamier'], 'ceamierr'	'creamier'

If they do, the algorithm computes the $QWERTY$ keyboard distance between each pair of characters in 'w' and 'word', stores the distances in an array *storage*, and computes the sum of the distances. The sum is then assigned to 'w'. This process is repeated for all words in 'words'. Finally, the algorithm returns the word from words that has the minimum distance to given 'word'. The Python code to auto-correct a given word is depicted in Code 5.10.

Code 5.10 Python code to auto-correct a given word

```
1   def keyword_distance():
2       # The order of letters in each line of usual keywords
3       top = {c: (0, i) for (i, c) in enumerate("qwertyuiop")}
4       mid = {c: (1, i) for (i, c) in enumerate("asdfghjkl")}
5       bot = {c: (2, i) for (i, c) in enumerate("zxcvbnm")}
6       # To update the dictionary with more than one argument
7       keys = dict(top, **mid, **bot)
8       dist = dict()
9       lows = "abcdefghijklmnopqrstuvwxyz"
10      # Computing the distance
11      for cc1 in lows:
12          for cc2 in lows:
13              (r1, c1) = keys[cc1]
14              (r2, c2) = keys[cc2]
15              dist[(cc1, cc2)] = \
16                  (abs(r2 - r1) + abs(c2 - c1)) ** 2
17      return dist
18  dist=keyword_distance()
19  # calls keyword_distance
20  '''
21  By using this function, there is no need
22  to compute the distance among all pairs each time.
23  '''
24  def ds(c1, c2):
25      return dist[(c1, c2)]
26  def autocorrectr_word(words,word):
27      '''
28      Extract words that has same
```

```
29          length with given word
30          '''
31          '''
32          strip: Removes the leading spaces
33          '''
34          '''
35          dict.fromkeys: Assigns a value to each
36          object in the sequence
37          '''
38       dists = dict.fromkeys\
39       ([x.strip() for x in words if len(x) == len(word)], 0)
40       for w in dists.keys():
41          '''
42          Store difference of current
43          word and given word
44          '''
45          storage=[]
46          for i in range(len(word)):
47              storage.append(ds(w[i], word[i]))
48          dists[w] = sum(storage)
49       return min(dists, key=dists.get)
```

5.11 Correct Verb Form in Spanish

In this challenge, the task is to find the correct forms of the verb in the Spanish language. Write a function that as input, takes verb, subject and tense, and returns the correct forms of the verb in the Spanish language. For some inputs, the expected outputs are illustrated in Table 5.11.

Algorithm

There are 3 types of regular verbs in Spanish. Hence; the three conditions are considered and for each of them, the three if condition specifies the type of their tense. For considering the type of time and verb, a Python dictionary is made. The keys

Table 5.11 The expected outputs for certain inputs to find correct verb form in Spanish

Verb, subject, tense	Expected output
'ganar', 'ustedes', 'pretérito'	'ganaron'
'escribir', 'ellos', 'imperfecto'	'escribían'
'tomar', 'nosotros', 'futuro'	'tomaremos'

in the dictionary indicate the subjects, and values indicate the values that are to be added to the given verb. In a simple statement, if the subject is found, add the value to the given word. The Python code to find the correct form of a verb in the Spanish language is depicted in Code 5.11.

Code 5.11 Python code to find the correct form of a verb into the Spanish language

```
1   def Correct_verb_form_in_Spanish(verb, subject, tense):
2       '''
3       A dictionary to store the letters
4       that will be added to the verb.
5       The key in di are subjects, and
6       valuse are objects that are to
7       be added to the verbs
8       '''
9       di = {}
10      # The first regular verbs
11      if verb[-2::] == 'ar':
12          # if the tense is present
13          if tense == 'presente':
14              di ={'yo': 'o', 'tú': 'as',
15              'él': 'a', 'ella':
16              'a', 'usted': 'a',
17              'nosotros': 'amos',
18              'nosotras': 'amos',
19              'vosotros': 'áis',
20              'vosotras': 'áis',
21              'ellos': 'an',
22              'ellas': 'an',
23              'ustedes': 'an'
24              }
25
26          elif tense == 'pretérito':
27              di = {'yo': 'é', 'tú': 'aste',
28              'él': 'ó', 'ella': 'ó',
29              'usted': 'ó',
30              'nosotros': 'amos',
31              'nosotras': 'amos',
32              'vosotros': 'asteis',
33              'vosotras': 'asteis',
34              'ellos': 'aron',
35              'ellas': 'aron',
36              'ustedes': 'aron'
37              }
38          elif tense == 'imperfecto':
39              di = {'yo': 'aba', 'tú': 'abas',
```

5.11 Correct Verb Form in Spanish

```
40            'él': 'aba', 'ella': 'aba',
41            'usted': 'aba',
42            'nosotros': 'ábamos',
43            'nosotras': 'ábamos',
44            'vosotros': 'abais',
45            'vosotras': 'abais',
46            'ellos': 'aban',
47            'ellas': 'aban',
48            'ustedes': 'aban'
49        }
50        '''
51    Repeat the strategy above for
52    the seond regular verbs
53    '''
54    elif verb[-2::] == 'er':
55        if tense == 'presente':
56            di = {'yo': 'o', 'tú': 'es',
57        'él': 'e', 'ella': 'e',
58        'usted': 'e',
59        'nosotros': 'emos',
60        'nosotras': 'emos',
61        'vosotros': 'éis',
62        'vosotras': 'éis',
63        'ellos': 'en',
64        'ellas': 'en', 'ustedes': 'en'
65        }
66        elif tense == 'pretérito':
67            di = {'yo': 'í', 'tú': 'iste',
68        'él': 'ió', 'ella': 'ió',
69        'usted': 'ió',
70        'nosotros': 'imos',
71        'nosotras': 'imos',
72        'vosotros': 'isteis',
73        'vosotras': 'isteis',
74        'ellos': 'ieron',
75        'ellas': 'ieron',
76        'ustedes': 'ieron'
77        }
78        elif tense == 'imperfecto':
79            di = {'yo': 'ía', 'tú': 'ías',
80        'él': 'ía', 'ella': 'ía',
81        'usted': 'ía',
82        'nosotros': 'íamos',
83        'nosotras': 'íamos',
84        'vosotros': 'íais',
```

```
85              'vosotras': 'íais',
86              'ellos': 'ían',
87              'ellas': 'ían', 'ustedes': 'ían'
88          }
89          '''
90      Repeat the strategy above for
91      the third regular verbs
92      '''
93      elif verb[-2::] == 'ir':
94          if tense == 'presente':
95              di = {'yo': 'o', 'tú': 'es',
96              'él': 'e', 'ella': 'e', 'usted': 'e',
97              'nosotros': 'imos',
98              'nosotras': 'imos',
99              'vosotros': 'ís',
100             'vosotras': 'ís',
101             'ellos': 'en',
102             'ellas': 'en', 'ustedes': 'en'
103         }
104         elif tense == 'pretérito':
105             di = {'yo': 'í', 'tú': 'iste',
106             'él': 'ió', 'ella': 'ió',
107             'usted': 'ió',
108             'nosotros': 'imos',
109             'nosotras': 'imos',
110             'vosotros': 'isteis',
111             'vosotras': 'isteis',
112             'ellos': 'ieron', 'ellas': 'ieron',
113             'ustedes': 'ieron'
114         }
115         elif tense == 'imperfecto':
116             di = {'yo': 'ía', 'tú': 'ías',
117             'él': 'ía', 'ella': 'ía',
118             'usted': 'ía',
119             'nosotros': 'íamos',
120             'nosotras': 'íamos',
121             'vosotros': 'íais',
122             'vosotras': 'íais',
123             'ellos': 'ían',
124             'ellas': 'ían',
125             'ustedes': 'ían'
126         }
127     '''
128     If the tense is future tense, it
129     doesn't matter what kind of verb it is
```

```
130        '''
131        if tense == 'futuro':
132            di = {'yo': 'é', 'tú': 'ás',
133            'él': 'á', 'ella': 'á',
134            'usted': 'á',
135            'nosotros': 'emos',
136            'nosotras': 'emos',
137            'vosotros': 'éis',
138            'vosotras': 'éis',
139            'ellos': 'án', 'ellas': 'án',
140            'ustedes': 'án'
141            }
142            '''
143            If the tense is future, then the
144            last two letters of the verb are not removed.
145            '''
146            return verb + di[subject]
147        '''
148        If the tense is not future, then the
149        last two letters of the verb are  removed.
150        '''
151        return verb[:len(verb) - 2:] + di[subject]
```

5.12 Subsequent Letters

A substring is a string whose order is the same as the original string, while a subsequent is a string in which the order of the letters is not necessarily the same as the original string. For example, '*tho*' is a substring of *python*, while **th**e*a*n**o** is a subsequent of *python*. Write a function that as input, takes *letter* as a list of letters, and *words* as a list of words, and returns all words that have the subsequent *letter*. For some inputs, the expected outputs are illustrated in Table 5.12.

Table 5.12 The expected outputs for certain inputs for finding the words that have the same subsequent letters

Words, letter	Expected output
['suits', 'refluxed', 'trip', 'refluxing', 'retroflux'], reflux	['refluxed', 'refluxing', 'retroflux']
['vasomotor', 'bathythermogram', 'benzhydroxamic', 'dialer'], byoam	['bathythermogram', 'benzhydroxamic']

Algorithm

It is used a combination of nested loops and string manipulation to find the words that have the same subsequent letters. The algorithm loops over each word in the input list and checks if the word is at least as long as the letters we're looking for. If so, the word is converted to a list of characters, and another loop is used to check if each letter we're looking for is present in the word. If all the letters are found in the word, the word is added to a list of matching words. Finally, the algorithm returns the list of matching words. The Python code to find all words that have the subsequent letters is depicted in Code 5.12.

Code 5.12 Python code to find all words that have the subsequent letters

```
1  def subsequent_letters(words, letters):
2      '''Get the length of the letters
3      we are looking for
4      '''
5      len_let = len(letters)
6      '''
7      Initialize an empty list
8      to store the matching words
9      '''
10     ST = []
11
12     # Loop over each word in the input list
13     for term in words:
14         '''
15         Check if the word is at least as long as
16         the letters we're looking for
17         '''
18         if len(term) >= len(letters):
19             # Convert the word to a list of characters
20             indterm = list(term)
21             '''
22             Initialize a counter and index for
23             tracking the letters we've found
24             '''
25             counter = 0
26             CurrentIndex = -1
27
28             '''
29             loop over each letter we're looking for
30             '''
31             for let in letters:
32                 '''
33                 Check if the letter is in the remaining
```

```
34                    characters of the word
35                    '''
36                    if let in indterm[CurrentIndex+1:]:
37                        '''
38                        If the letter is found,
39                        update the index and counter
40                        '''
41                        indexlet = \
42                        indterm.index(let, CurrentIndex+1)
43                        CurrentIndex = indexlet
44                        counter += 1
45
46                        '''
47                        'If we have found all the
48                        letters, add the word to
49                        the list and break the loop
50                        '''
51                        if counter == len_let:
52                            ST.append(term)
53                            break
54
55     # Return the list of matching words
56     return ST
```

5.13 Possible Words from a Text Corpus

In this challenge, for a given string pattern, all words that include the pattern must be found. For example, if $p = ***b*ls$ is the given pattern, '*jumbals*' and '*verbals*' are words that include the letters of the pattern. Write a function that as input, takes a string pattern, and returns all words that include the pattern. For some inputs, the expected outputs are illustrated in Table 5.13.

Table 5.13 The expected outputs for certain inputs for extracting possible words from a text corpus

Words, pattern	Expected output
['pepsi', 'fissury', 'dark', 'missary', 'missort'] , '*iss*r*'	['fissury', 'missary', 'missort']
['havened', 'car', 'hoveled', 'people', 'hovered'] 'h*ve*ed' ,	['havened', 'hoveled', 'hovered']

Algorithm

Let *words* be a list of words given, and *pattern* be a string that includes the letters. For each *word* in *words*, it encodes all letters of *word* and *pattern*. It checks if all letters in *pattern* are in *word*, and their encoding is the same, store *word* into the array *result*, where *result* is all words that include *pattern*.

The Python code to find all words that have the given string pattern is depicted in Code 5.13.

Code 5.13 Python code to find all words that have the given string pattern

```
1   def encode(word, pattern):
2       '''
3       In python, encoding of * is 42
4       '''
5       encoding_pattern = list(pattern.encode())
6       encoding_word = list(word.encode())
7       pattern_filter = \
8       [x for x in encoding_pattern if x != 42]
9       encoded_word = []
10      '''
11      It is enough if the letters in word be in pattern.
12      It doesn't matter what the rest of the letters are.
13      '''
14      for int_ in encoding_word:
15          if int_ in pattern_filter:
16              encoded_word.append(int_)
17          else:
18              encoded_word.append(42)
19      return encoding_pattern, encoded_word
20
21  def possible_words(words, pattern):
22      '''
23      Encode the pattern and each word, and
24      if their encoding is equal, consider
25      the word as a possible one.
26      '''
27      result = []
28      pattern_encoding = encode('', pattern)[0]
29      for word in words:
30          '''
31          To avoid checking all words,
32          if any two words have the same length, then
33          they are encoded.
34          '''
35          if len(word) == len(pattern):
```

5.13 Possible Words from a Text Corpus

```
36              word_encoding = encode(word, pattern)[1]
37              if word_encoding == pattern_encoding:
38                  result.append(word)
39      return result
```

Chapter 6
Game

This chapter discusses 14 game-based problems, which are explained through examples and subsequently programmed in Python. The problems are as follows:

1. Determining the card that leads the player to win the game
2. Hand shape in Bridge game
3. Obtaining the abbreviations in the contract bridge game
4. Same hand shape distribution in a card game
5. The number of rounds in a given permutation
6. Reaching stable state in candy share
7. African Oware game
8. The number of Safe Squares Rooks of Chessboard
9. The number of Safe Squares not threatened by bishops
10. Reaching Knight jump
11. Capturing maximum checkers in chessboard
12. The number of safe squares for friend pieces in a chessboard
13. Crag Score
14. Optimal crag score with multiple rolls

6.1 Winner in Card Game

In playing card games, there is a game with four players each one has one card in form (rank, suit), where a suit is a category of the cards. If a trump card is included in the game, so trump is one of the elements in c, $c = \{`clubs', `spades', `hearts', `diamonds'\}$. The order of ranks is $`ace' > `king' > `queen' > `jack' > 10 > 9 > 8 > 7 > 6 > 5 > 4 > 3 > 2 > 1$. With some tricks, one can win the game. The rules for the tricking are as the followings:

1. If a trump card is given, the first trump card with the highest rank is selected, and its player wins the game.

Table 6.1 The expected outputs for certain inputs for determining the winner in the card game

Cards, trump	Expected output
[('three', 'clubs'), ('king', 'spades'), ('queen', 'clubs'), ('jack', 'hearts')] , queen	('queen', 'clubs')
[('four', 'diamonds'), ('ace', 'clubs'), ('four', 'spades'), ('king', 'clubs')],None	('four', 'diamonds')
[('eight', 'hearts'), ('three', 'diamonds'), ('nine', 'spades'), ('queen', 'hearts')] , None	('queen','hearts')

2. If a trump card is not given, the first card with the highest rank is selected, and its player wins the game.

For example, if cards=[('three', 'clubs'), ('king', 'spades'), ('queen', 'clubs'), ('jack', 'hearts')], and $trump = queen$, the first player with the card ('$queen'$, '$clubs'$) is won. Although ('$three'$, '$clubs'$) is the first card, but ('$queen'$, '$clubs'$) has more rank. In another example, if cards=[('four', 'diamonds'), ('ace', 'clubs'), ('four', 'spades'), ('king', 'clubs')], and $trump = None$, the first player with the card ('$four'$, '$diamonds'$) is won. For some inputs, the expected outputs are illustrated in Table 6.1.

Write a function that as input, takes a list of cards, and returns, the card that leads the player to win the game.

Algorithm

For each suit of spades, hearts, diamonds, and clubs, the program stores their respective ranks and sorts them using the bubble sort algorithm according to a set of rules. In the subsequent step, the program selects the card with the highest rank based on the following criteria: If a trump card is given as input, it selects the first trump card with the highest rank. If a trump card is not given in the input, it selects the first card with the highest rank. The Python code to find and return the card that leads the player to win the game is depicted in Code 6.1.

Code 6.1 Python code to find and return the card that lead the player to win the game

```
1  def bubble_sort(arr):
2      for i in range(len(arr) - 1):
3          for j in range(0, (len(arr) - i) - 1):
4              if arr[j] > arr[j + 1]:
5                  # Swap if it is needed
6                  arr[j], arr[j + 1] = arr[j + 1], arr[j]
7      return arr
8  def Winner_in_Card_Game(cards, trump=None):
9      names =[x[1] for x in cards]
10     '''
```

6.1 Winner in Card Game

```
11      The arrays are defined for
12      storing the ranks if the trumps.
13      '''
14      spades = []
15      hearts = []
16      diamonds = []
17      clubs = []
18      '''
19      dicnames is defined for finding
20      the ranks via their numerical ranks
21      '''
22      dicnames = {1: 'one', 2: 'two', 3: 'three', 4: 'four', 5:
23      'five', 6: 'six', 7: 'seven',
24      8: 'eight', 9: 'nine', 10: 'ten', 50: 'ace',
25      40: 'king', 30: 'queen', 20: 'jack'}
26      for item in cards:
27          if item[1] == 'spades':
28              if item[0] == 'ace':
29                  spades.append(50)
30              elif item[0] == 'king':
31                  spades.append(40)
32              elif item[0] == 'queen':
33                  spades.append(30)
34              elif item[0] == 'jack':
35                  spades.append(20)
36              elif item[0] == 'one':
37                  spades.append(1)
38              elif item[0] == 'two':
39                  spades.append(2)
40              elif item[0] == 'three':
41                  spades.append(3)
42              elif item[0] == 'four':
43                  spades.append(4)
44              elif item[0] == 'five':
45                  spades.append(5)
46              elif item[0] == 'six':
47                  spades.append(6)
48              elif item[0] == 'seven':
49                  spades.append(7)
50              elif item[0] == 'eight':
51                  spades.append(8)
52              elif item[0] == 'nine':
53                  spades.append(9)
54              elif item[0] == 'ten':
55                  spades.append(10)
```

```
56
57              elif item[1] == 'hearts':
58                  if item[0] == 'ace':
59                      hearts.append(50)
60                  elif item[0] == 'king':
61                      hearts.append(40)
62                  elif item[0] == 'queen':
63                      hearts.append(30)
64                  elif item[0] == 'jack':
65                      hearts.append(20)
66                  elif item[0] == 'one':
67                      hearts.append(1)
68                  elif item[0] == 'two':
69                      hearts.append(2)
70                  elif item[0] == 'three':
71                      hearts.append(3)
72                  elif item[0] == 'four':
73                      hearts.append(4)
74                  elif item[0] == 'five':
75                      hearts.append(5)
76                  elif item[0] == 'six':
77                      hearts.append(6)
78                  elif item[0] == 'seven':
79                      hearts.append(7)
80                  elif item[0] == 'eight':
81                      hearts.append(8)
82                  elif item[0] == 'nine':
83                      hearts.append(9)
84                  elif item[0] == 'ten':
85                      hearts.append(10)
86              elif item[1] == 'diamonds':
87                  if item[0] == 'ace':
88                      diamonds.append(50)
89                  elif item[0] == 'king':
90                      diamonds.append(40)
91                  elif item[0] == 'queen':
92                      diamonds.append(30)
93                  elif item[0] == 'jack':
94                      diamonds.append(20)
95                  elif item[0] == 'one':
96                      diamonds.append(1)
97                  elif item[0] == 'two':
98                      diamonds.append(2)
99                  elif item[0] == 'three':
100                     diamonds.append(3)
```

6.1 Winner in Card Game

```
101                elif item[0] == 'four':
102                    diamonds.append(4)
103                elif item[0] == 'five':
104                    diamonds.append(5)
105                elif item[0] == 'six':
106                    diamonds.append(6)
107                elif item[0] == 'seven':
108                    diamonds.append(7)
109                elif item[0] == 'eight':
110                    diamonds.append(8)
111                elif item[0] == 'nine':
112                    diamonds.append(9)
113                elif item[0] == 'ten':
114                    diamonds.append(10)
115            elif item[1] == 'clubs':
116                if item[0] == 'ace':
117                    clubs.append(50)
118                elif item[0] == 'king':
119                    clubs.append(40)
120                elif item[0] == 'queen':
121                    clubs.append(30)
122                elif item[0] == 'jack':
123                    clubs.append(20)
124                elif item[0] == 'one':
125                    clubs.append(1)
126                elif item[0] == 'two':
127                    clubs.append(2)
128                elif item[0] == 'three':
129                    clubs.append(3)
130                elif item[0] == 'four':
131                    clubs.append(4)
132                elif item[0] == 'five':
133                    clubs.append(5)
134                elif item[0] == 'six':
135                    clubs.append(6)
136                elif item[0] == 'seven':
137                    clubs.append(7)
138                elif item[0] == 'eight':
139                    clubs.append(8)
140                elif item[0] == 'nine':
141                    clubs.append(9)
142                elif item[0] == 'ten':
143                    clubs.append(10)
144    bubble_sort(spades)
145    bubble_sort(hearts)
```

```
146         bubble_sort(diamonds)
147         bubble_sort(clubs)
148         '''
149         In python, eval is used to evaluate
150         an expression. In a simple statement,
151         it takes an string and returns the
152         true value of an object.
153         t= eval('9')-----9, t=9, t is an int.
154         t=eval('2*x+1')----- t=13
155         '''
156         # Use the mentioned rules to determine the winner
157         '''
158         eval(cards[0][1]) returns the true
159         value spades or hearts or diamonds
160         or clubs as a string, [-1] returns
161         the largest number, as the array is
162         sorted in an ascending order.
163         '''
164         if trump not in names:
165             return dicnames[eval(cards[0][1])[-1]],cards[0][1]
166         # If a trump is given in the input
167         elif trump in names:
168             '''
169             eval(trump) returns the true value
170             spades or hearts or diamonds or clubs
171             as a string, [-1] returns the largest number,
172             as the array is sorted in an ascending order.
173             '''
174             return dicnames[eval(trump)[-1]],trump
```

6.2 Hand Shape in Bridge Game

In the card game bridge, each player has thirteen cards. In this challenge, the number of occurrences of each card must be returned, where the order of the cards is $\{'spades', 'hearts', 'diamonds', 'clubs'\}$. Write a function that as input takes the thirteen cards, and returns the number of the occurrences of each trump card in order $\{'spades', 'hearts', 'diamonds', 'clubs'\}$. For one inputs, the expected output is illustrated in Table 6.2.

6.3 Contract Bridge Game

Table 6.2 The expected outputs for certain inputs for hand shape in bridge game

Cards	Expected output
[('jack', 'diamonds'), ('jack', 'hearts'), ('seven', 'clubs'), ('five', 'clubs'), ('ace', 'diamonds'), ('three', 'clubs'), ('four', 'spades'), ('three', 'spades'), ('eight', 'spades'), ('ace', 'hearts'), ('five', 'diamonds'), ('two', 'clubs'), ('queen', 'diamonds')]	[3, 2, 4, 4]

Algorithm

To determine the shape of a hand in a game of bridge, it is necessary to iterate through the provided cards and count the number of objects in the set c, where c represents the four suits: spades, hearts, diamonds, and clubs. The Python code that accomplishes this task and returns the hand shape is presented in Code 6.2

Code 6.2 Python code to find and return the shape of a hand in card game bridge

```
def Hand_Shape_in_Bridge_Game(hand):
    cards = []
    countinglist = []
    for idx, shapes in hand:
        # shapes is the card names
        cards.append(shapes)
    spades = cards.count('spades')
    hearts = cards.count('hearts')
    diamonds = cards.count('diamonds')
    clubs = cards.count('clubs')
    countinglist=[spades, hearts, diamonds, clubs]
    return countinglist
```

6.3 Contract Bridge Game

In the game of contract bridge, each player is dealt a hand of thirteen cards. The ranks of these cards can be abbreviated to create a more visually appealing display. The objective is to find the appropriate abbreviations for each rank, with the order of cards always being spades, hearts, diamonds, and clubs. The abbreviations are as follows: 'A' for Ace, 'K' for King, 'Q' for Queen, 'J' for Jack, and 'x' for all other ranks. To accomplish this task, a function must be written that takes a list of thirteen cards as input and returns the abbreviated form in the order 'AQKJ'. The 'x' can be placed in any position within the abbreviation order, and if a suit is empty, a '-' is to be placed in the corresponding position. For one inputs, the expected output is illustrated in Table 6.3.

Table 6.3 The expected outputs for certain inputs for hand shape in contract bridge game

Cards	Expected output
[('three', 'clubs'), ('ten', 'spades'), ('jack', 'hearts'), ('five', 'hearts'), ('jack', 'clubs'), ('two', 'diamonds'), ('eight', 'hearts'), ('eight', 'clubs'), ('three', 'spades'), ('ace', 'hearts'), ('jack', 'spades'), ('king', 'diamonds'), ('six', 'hearts')]	'JxxAJxxxKxJxx'

Algorithm

The algorithm first encodes the ranks of the cards in each suit as numeric values, with 'ace' being assigned the lowest value (0) and 'x' being assigned the highest value (4). Then, the algorithm uses the bubble sort algorithm to sort the cards in each suit in ascending order based on their encoded rank. Finally, the algorithm replaces the encoded values with the corresponding card names, such as 'A' for 'ace', 'K' for 'king', and so on, and returns the sorted cards for each suit as a string. The Python code to find and return the abbreviations in the contract bridge game is depicted in Code 6.3.

Code 6.3 Python code to find and return the abbreviations in contract bridge game is depicted in

```
1   def bubble_sort(arr):
2       for i in range(len(arr) - 1):
3           for j in range(0, (len(arr) - i) - 1):
4               if arr[j] > arr[j + 1]:
5                   # Swap if it is needed
6                   arr[j], arr[j + 1] = arr[j + 1], arr[j]
7       return arr
8   def contract_bridge_game(hand):
9
10      spades = []
11      hearts = []
12      diamonds = []
13      clubs = []
14      '''
15      It encodes the ranks for being
16      sortable in the next steps.
17      '''
18      for item in hand:
19          if item[1] == 'spades':
20              if item[0] == 'ace':
21                  spades.append('0')
22              elif item[0] == 'king':
23                  spades.append('1')
```

6.3 Contract Bridge Game

```
24                elif item[0] == 'queen':
25                    spades.append('2')
26                elif item[0] == 'jack':
27                    spades.append('3')
28                else:
29                    spades.append('4')
30            elif item[1] == 'hearts':
31                if item[0] == 'ace':
32                    hearts.append('0')
33                elif item[0] == 'king':
34                    hearts.append('1')
35                elif item[0] == 'queen':
36                    hearts.append('2')
37                elif item[0] == 'jack':
38                    hearts.append('3')
39                else:
40                    hearts.append('4')
41            elif item[1] == 'diamonds':
42                if item[0] == 'ace':
43                    diamonds.append('0')
44                elif item[0] == 'king':
45                    diamonds.append('1')
46                elif item[0] == 'queen':
47                    diamonds.append('2')
48                elif item[0] == 'jack':
49                    diamonds.append('3')
50                else:
51                    diamonds.append('4')
52            elif item[1] == 'clubs':
53                if item[0] == 'ace':
54                    clubs.append('0')
55                elif item[0] == 'king':
56                    clubs.append('1')
57                elif item[0] == 'queen':
58                    clubs.append('2')
59                elif item[0] == 'jack':
60                    clubs.append('3')
61                else:
62                    clubs.append('4')
63        # specifying the absents
64        if len(spades) == 0:
65            spades.append('-')
66        if len(hearts) == 0:
67            hearts.append('-')
68        if len(diamonds) == 0:
```

```
69          diamonds.append('–')
70      if len(clubs) == 0:
71          clubs.append('–')
72      bubble_sort(spades)
73      bubble_sort(hearts)
74      bubble_sort(diamonds)
75      bubble_sort(clubs)
76      # replacing the encodes with the wanted names
77      output = ''.join(spades)+'    '+\
78          '    '.join(hearts)+'    '+\
79          '    '.join(diamonds)+\
80          '    '+'    '.join(clubs)
81      output = output.replace('0', 'A')\
82          .replace('1', 'K').replace(
83          '2', 'Q').replace('3', 'J')\
84          .replace('4', 'x')
85      return output
```

6.4 Same Hand Shape Distribution

In this challenge, the two hands that have different permutations can be considered as the same. For example, [6, 3, 2, 2] and [2, 3, 6, 2] have the same shape. Write a function that as input takes a list of hands, and returns a list of tuples that contain the same shapes, where the tuples are sorted in descending order, and the list of tuples is sorted in ascending order. For one input, the expected output is illustrated in Table 6.4.

Algorithm

By using a linear search, the found suits are enumerated and inserted into an array, and it applies the wanted sorts to the considered array. The Python code to find and return the shapes that contain the same distributions is depicted in Code 6.4.

Code 6.4 Python code to find and return the shape that contains the same distributions

```
1  def insertion_sort_descending(arr):
2      for i in range(1, len(arr)):
3          key = arr[i]
4          j = i - 1
5          while j >= 0 and key > arr[j]:
6              arr[j + 1] = arr[j]
7              j -= 1
8          arr[j + 1] = key
```

6.4 Same Hand Shape Distribution

Table 6.4 The expected output for certain input for same hand shape distribution

Cards, trump	Expected output
[('two', 'hearts'), ('nine', 'spades'), ('two', 'clubs'), ('eight', 'diamonds'), ('queen', 'diamonds'), ('ace', 'clubs'), ('six', 'diamonds'), ('queen', 'hearts'), ('three', 'hearts'), ('queen', 'clubs'), ('ten', 'spades'), ('nine', 'clubs'), ('six', 'spades')], [('five', 'diamonds'), ('jack', 'hearts'), ('three', 'spades'), ('king', 'clubs'), ('two', 'spades'), ('king', 'spades'), ('jack', 'spades'), ('jack', 'diamonds'), ('seven', 'clubs'), ('nine', 'hearts'), ('two', 'hearts'), ('king', 'hearts'), ('eight', 'hearts')], [('queen', 'clubs'), ('jack', 'clubs'), ('two', 'diamonds'), ('nine', 'spades'), ('six', 'clubs'), ('ace', 'clubs'), ('ten', 'diamonds'), ('nine', 'diamonds'), ('three', 'clubs'), ('five', 'hearts'), ('eight', 'spades'), ('six', 'spades'), ('eight', 'hearts')], [('seven', 'spades'), ('ten', 'spades'), ('two', 'hearts'), ('two', 'clubs'), ('jack', 'spades'), ('five', 'spades'), ('queen', 'clubs'), ('king', 'hearts'), ('king', 'clubs'), ('seven', 'diamonds'), ('eight', 'clubs'), ('queen', 'spades'), ('four', 'spades')]]	[((4, 3, 3, 3), 1), ((5, 3, 3, 2), 1), ((5, 4, 2, 2), 1), ((6,4,2,1), 1)]

```
9        return arr
10   def bubble_sort(arr):
11        for i in range(len(arr) - 1):
12            for j in range(0, (len(arr) - i) - 1):
13                if arr[j] > arr[j + 1]:
14                    # Swap if it is needed
15                    arr[j], arr[j + 1] = arr[j + 1], arr[j]
16        return arr
17   def hand_shape_distribution(hands):
18        arr = []
19        arr1 = []
20        for i in range(len(hands)):
21            hand_list = [0, 0, 0, 0]
22            # Counting the number of occurrences
23            for j in hands[i]:
24                if j[1] == 'spades':
25                    hand_list[0] += 1
26                elif j[1] == 'hearts':
27                    hand_list[1] += 1
28                elif j[1] == 'diamonds':
```

```
29                    hand_list[2] += 1
30                elif j[1] == 'clubs':
31                    hand_list[3] += 1
32            hand_list= insertion_sort_descending( hand_list)
33            arr.append(hand_list)
34        for i in arr:
35            arr1.append([i, arr.count(i)])
36        # To remove the hands that have not the same shape
37        temp_list = []
38        for i in arr1:
39            if i not in temp_list:
40                temp_list.append(i)
41        arr1 = temp_list
42        #Converting to tuple
43        for j in range(len(arr1)):
44            arr1[j][0] = tuple(arr1[j][0])
45            arr1[j] = tuple(arr1[j])
46        # Sorting in ascending order
47        arr1=bubble_sort(arr1)
48        return arr1
```

6.5 Number Round Counter

In this challenge, you are given an array of numbers which may be either sorted or unsorted. The array is filled with numbers from a permutation of 0 to $n - 1$. The objective of this challenge is to compute the number of rounds required to collect the numbers in ascending order. In the first round, the numbers are considered in ascending order. In the subsequent rounds, the numbers that have not been considered in the previous rounds are taken into account, again in ascending order. This process is repeated until all the numbers in ascending order have been visited. For example, let us consider the given list $l = [0, 4, 3, 5, 2, 1]$. In the first round, 0 and 1 are in ascending order, in the second round, just 2 is in ascending order, in the third round, 3 is in ascending order, and in the fourth round, 4 and 5 are in ascending order. Hence, there are four rounds for list L. Write a function that as input takes a permutation of numbers from 0 to $n - 1$, and returns the numbers of rounds. For some inputs, the expected outputs are illustrated in Table 6.5.

Algorithm

To solve this challenge the method of inverse permutation is used. Let *perm* be the given array, and *inverseperm* is an array equal to size *perm* that is filled with

6.5 Number Round Counter

Table 6.5 The expected outputs for certain inputs for counting the rounds in an array

Perm	Expected output
[0, 1, 5,4,2, 3]	3
[0, 1,2,4,3,5]	2
[4, 11, 2,8, 6, 9, 5, 3,0 ,10, 1, 12,7]	6
[0,1,4,3,2]	3

zeros initially, $inverseperm$ is utilized to inverse the given permutation. For each i in $perm$, statement $inverseperm[perm[i]] = i$ inverse the permutations. To count the number of rounds $rounds$ is defined. Initially, $rounds$ is set to 1, as there is always a sequence of numbers in ascending order that the sequence, where the cardinality of the sequence is at least one. Let n be the cardinality of $inverseperm$, for $k = 1$ to n, if $inverseperm[k] < inverseperm[k-1]$, so k is on the left side $k-1$, and $rounds$ must be incremented by one. The Python code to compute and return the number of rounds in which the numbers are collected in ascending order is depicted in Code 6.5.

Code 6.5 Python code to compute the number of rounds to collect the numbers in ascending order

```
1   def Number_round_counter(perm):
2       '''
3       perm is an array that consisted
4       of numbers from 0 to n-1
5       '''
6       # To store the number of rounds
7       rounds = 1
8       '''
9       Create an array with the
10      size the of perm array
11      '''
12      inverse_perm = [0] * len(perm)
13      for i in range(len(perm)):
14          '''
15          Inversing the the perm array
16          '''
17          inverse_perm[perm[i]] = i
18      for k in range(1, len(inverse_perm)):
19          '''
20          In the inverse_perm array,
21          if each element is smaller than
22          its previous element,
23          it means that this number is on
24          the left side of its previous
25          number in the perm list, as a result,
```

```
26                  the round variable must be incremented by one.
27              '''
28              if inverse_perm[k] < inverse_perm[k - 1]:
29                  rounds += 1
30      return rounds
```

6.6 Reaching Stable State in Candy Share

A group of children is seated around a circular table with candies placed in front of them. The candies are indistinguishable from each other. In order to synchronize the children, their teacher rings a bell. When the bell rings, each child who has at least two candies must send one candy to the child on their left and one to the child on their right. Children with one or zero candies are not required to take any action. The application of the pigeonhole principle demonstrates that when there are more candies than children, the program will never terminate. Write a function that takes the initial distribution of candies as its input and returns the number of steps required to reach a stable state. A stable state is one in which it is not possible to transfer any more candies to the right or left. The expected outputs for certain inputs are shown in Table 6.6.

Algorithm

It recursively applies a set of rules to the input 'candies' list until a stable state is reached, where no child has more than one candy. At each iteration, the algorithm identifies a child with more than two candies and transfers one candy to each of its adjacent children. This process is repeated until all children have at most one candy, at which point the algorithm terminates and returns the number of states it took to reach the stable state. The algorithm steps are outlined in detail as follows:

1. It takes two parameters: 'candies': a list of integers representing the number of candies each child has. 'states' (optional): an integer representing the number of states that the program has reached a stable state (default is 0).

Table 6.6 The expected outputs for certain inputs for reaching stable state in candy share

Candies	Expected output
[0, 0, 0, 3]	1
[2, 0, 0, 1]	2
[2,0,0,0,0,0,4,0, 1]	8
[0,4,0,0,0,3,0, 1,0,0]	4

6.6 Reaching Stable State in Candy Share

2. Check if the base case is reached. If the number of children is 0 or 1, candies is equal to the total number of children, return the number of states.
3. Create a copy of the 'candies' list for the current round.
4. Iterate over each child in the 'candies' list.
5. Check if the current child has at least two candies. If so, remove two candies from the current child and give one candy to each of the neighboring children (i.e., the child to the left and the child to the right).
6. Recursively do the above steps with the updated 'candies' list and the incremented 'states'.
7. Return the final value of 'states' when the base case is reached.

The Python code to return the number of states reaching to a stable state in the candy share problem is depicted in Code 6.6.

Code 6.6 Python code to return the number of states reaching to a stable state in candy share problem

```
1  def candy_share(candies, states=0):
2      '''
3      For returning the number of
4      states that the program
5      have reached a stable state.
6      '''
7      # Base case: all children have at most 1 candy
8      if candies.count(0)+candies.count(1)==len(candies):
9          return states
10
11     # A copy of candies list for any round
12     previous_state=candies.copy()
13
14     '''
15     To apply the laws
16     of the game for every child
17     '''
18     for i in range(len(candies)):
19         '''
20         Only a child who has more
21         than two candies in each round
22         must transfer its candies to left
23         and right
24         '''
25         if previous_state[i] >= 2:
26             # Remove candies of the child
27             candies[i] -= 2
28
29             # Transfer them to left and right
30             candies[(i - 1) % len(candies)] += 1
```

```
31                candies[(i + 1) % len(candies)] += 1
32
33        # Recursive call with updated candies and states
34        return candy_share(candies, states+1)
```

6.7 Oware Game

Oware is an African board game with many variations its popular one is Mancala. In this game, there is a board with two rows, and in each row, some holes (house), and players perform 'sow' and can 'capture' seeds, where the seeds are in holes. In this challenge, there is a board with size $2n$, where the first n represents the houses of the first player, and the last n represents the houses of the opponent. The seeds are sowed and then are to be picked up, and the house captured has either two seeds or three seeds. For example, if the inputs are $board = [0, 2, 1, 2]$, and $house = 1$, output an array that shows the changes after picking up the seeds. The first phase is for sowing, where the sowing is started from position $house + 1$ to a value located in $board[house]$, and the value in $house$ position (index) of $board$ becomes zero, so $board = [0, 0, 1, 2]$. In the second position, one seed is sown, so $board = [0, 0, 2, 2]$, in the next step, one seed in the third position is sown so $board = [0, 0, 2, 3]$. The second and third positions are captured, as they are either two or three. Therefore the output is $[0, 0, 0, 0]$. Write a function that as input, takes a board, and returns an array that shows the changes after picking up the seeds. For some inputs, the expected outputs are illustrated in Table 6.7.

Algorithm

The algorithm uses a while loop and a for loop to iterate over the Oware board and sow the seeds to the houses and capture opponent's seeds if applicable.

1. It takes *board*: a list representing the current state of the Oware board and *house*: an integer representing the index of the house from which the player wants to start sowing seeds.

Table 6.7 The expected outputs for certain inputs for African oware game

Board,house	Expected output
[2, 1, 2,0],1	[2, 1, 2,0]
[1,4,5,6],1	[2, 0, 7, 7]
[7,7,7,68,0,1,0],3	[18, 18, 18, 0, 12, 13, 11]
[2, 0, 7, 7],1	[2, 0, 8, 7]

6.7 Oware Game

2. Determine the last index of the board by subtracting 1 from the length of the board list, denoted by *last_index*.
3. Store the number of seeds in the selected house in a variable called *seeds_to_sow*.
4. Set the current index (*current_index*) to the selected house plus 1.
5. Set the number of seeds in the selected house to 0.
6. While *seeds_to_sow* is greater than 0, do the following:

   a. If the current index goes beyond the last index, go back to the first index.
   b. If the current index is equal to house, skip the starting house.
   c. Add one seed to the current house.
   d. Reduce the number of seeds to sow by one.
   e. Move to the next house by adding 1 to the current index.
   a. Iterate through the opponent's houses starting from the last house that received a seed and moving backwards towards the opponent's store.
   b. For each opponent's house, do the following:
      i. If the house has 2 or 3 seeds, capture them and set the number of seeds in the house to 0.
      ii. Stop capturing if the opponent's house has more than 3 seeds.

7. Return the updated *board* list.

The Python code to returns an array that shows the changes after picking up the seed in the African oware game is depicted in Code 6.7.

Code 6.7 Python code to returns an array that shows the changes after picking up the seed in African oware game

```python
def oware_game(board, house):
    # Determine the last index of the board
    last_index = len(board) - 1
    # Store number of seeds in the selected house
    seeds_to_sow = board[house]
    # Set the current index to the selected house
    current_index = house + 1
    # Set number of seeds in the selected house to 0
    board[house] = 0
    # Distribute the seeds to the adjacent houses
    while seeds_to_sow > 0:
        '''
        If the current index goes
        beyond the last index,
        go back to the first index
        '''
        if current_index > last_index:
            current_index = 0
        # Skip the starting house
        if current_index == house:
            current_index += 1
```

```
22          # Add one seed to the current house
23          board[current_index] += 1
24          # Reduce the number of seeds to sow by one
25          seeds_to_sow -= 1
26          # Move to the next house
27          current_index += 1
28
29      # Capture opponent's seeds if applicable
30      end_index = current_index - 1
31      if end_index >= len(board) / 2:
32          for i in range(end_index, len(board) // 2-1, -1):
33              '''
34              If the opponent's house has 2 or 3 seeds,
35              capture them and set the number of seeds
36              in the house to 0
37              '''
38              if board[i] in (2, 3):
39                  board[i] = 0
40              '''
41              Stop capturing if the opponent's house
42              has more than 3 seeds
43              '''
44              else:
45                  break
46
47      # Return the updated board
48      return board
```

6.8 Safe Squares in Chessboard

The rook is a game piece in the game of chess, and each rook is represented by a two-tuple (row, column), where the rows and columns are numbered from 0 to $n-1$. The rooks have the ability to invade an entire row or column on the $n \times n$ chessboard. The objective of this challenge is to identify the set of safe squares on the chessboard where no two rooks can attack them. Write a function that as input, takes rooks as a list of tuples, and the size of the chessboard, and returns the number of safe squares. For some inputs, the expected outputs are illustrated in Table 6.8.

6.9 Safe Squares Not Threaten with Bishops

Table 6.8 The expected outputs for certain inputs for counting the safe squares in a chessboard

n,rooks	Expected output
10,[(2,3),(4,4)]	64
91,[(1,1),(4,4), (3, 5), (0, 7)]	7569
20,[(1,0),(3,6), (11, 5), (1, 2)]	272
7,[(0,2),(3,9), (3, 4)]	20

Algorithm

Let *rooks* be the given lists of rooks, and *n* be the size of the chessboard. For each *i* in *rooks*, it stores *i*[0] into an array $Unsaferow$, and stores *i*[1] into an array $Unsafecol$. In the next step, it obtains the safed rows and the safed columns by $safedrow = n - len(set(Unsafer))$, $safecol = n - len(set(Unsafecol))$, receptively. In the last step, the number of safed squares is obtained by $safedrow * safecol$.

The Python code for returning the number of safe squares in a chessboard is depicted in Code 6.8.

Code 6.8 Python code for returning the number of safe squares in a chessboard

```
1  def Safe_Squares_Rooks_of_Chessboard(n, rooks):
2      Unsafe_squares_in_rows = []
3      Unsafe_squares_in_columns = []
4      for i in rooks:
5          Unsafe_squares_in_rows.append(i[0])
6          Unsafe_squares_in_columns.append(i[1])
7      # to remove duplicated numbers, set is apllied
8      safed_rows = n - len(set(Unsafe_squares_in_rows))
9      # to remove duplicated numbers, set is apllied
10     safed_columns=n -len(set(Unsafe_squares_in_columns))
11     # Obtaining the safed_squares
12     safe = safed_rows * safed_columns
13     return safe
```

6.9 Safe Squares Not Threaten with Bishops

There is a $n \times n$ chessboard with some bishops pieces on it, where the rows and columns are numbered from zero to $n - 1$. The elephant pieces can only move diagonally. The bishops can threaten the squares of the chessboard. Write a function that as input takes bishops as a list of tuples, and the size of the chessboard and returns

Table 6.9 The expected outputs for certain inputs for counting the safe squares that are not threatened with bishops

n,bishops	Expected output
10,[(2,3),(4,4)]	68
91,[(1,1),(4,4), (3, 5), (0, 7)]	8002
20,[(1,0),(3,6), (11, 5), (1, 2)]	307
7,[(0,2),(3,9), (3, 4)]	29

the number of squares that are not threatened by any bishops. For some inputs, the expected outputs are illustrated in Table 6.9.

Algorithm

The algorithm takes as input the size of a chessboard 'n' and a list of bishops on the board 'bishops'. It calculates the number of safe squares on the board, that is, the number of squares that are not occupied by a bishop or threatened by a bishop on the same diagonal. The algorithm first initializes a variable 'safe_squares' to 0, which will be used to count the number of safe squares. It then creates two sets 'diagonal1' and 'diagonal2' to store the occupied squares on each diagonal of the chessboard.

For each bishop on the board, the algorithm calculates the indices of the squares on the two diagonals that are threatened by the bishop, and adds them to the respective sets 'diagonal1' and 'diagonal2'. Finally, the algorithm iterates over all squares on the board and checks if each square is occupied or threatened by a bishop on the same diagonal. If a square is not occupied or threatened, the algorithm increments the count of safe squares by 1. The algorithm returns the count of safe squares as the output. The Python code for returning the number of safe squares in a chessboard with bishops is depicted in Code 6.8.

Code 6.9 Python code for returning the number of safe squares in a chessboard with bishops

```
1   def Safe_Squares_not_Threaten_with_Bishops(n, bishops):
2       safe_squares = 0
3
4       # Store sets of occupied squares on each diagonal
5       # stores occupied squares on diagonal 1
6       diagonal1 = set()
7       # stores occupied squares on diagonal 2
8       diagonal2 = set()
9
10      # Iterate over bishops and update diagonal sets
11      for bishop in bishops:
12          # add square to diagonal 1
13          diagonal1.add(bishop[0] + bishop[1])
```

Table 6.10 The expected outputs for certain inputs to reach knight jump

Knight, start, end	Expected output
(2,1,7),(3,5,9),(8,11,13)	False
(3,4,2),(3,5,9),(1,7,9)	False
(2, 1), (12, 10),(11, 12)	True
(10, 5, 1), (20, 11, 16), (10, 12, 11)	True

```
14              # add square to diagonal 2
15              diagonal2.add(bishop[0] - bishop[1])
16
17          # Count safe squares
18          for i in range(n):
19              for j in range(n):
20                  if i + j not in diagonal1 \
21                  and i - j not in diagonal2:
22                      # increment safe squares count
23                      safe_squares += 1
24          return safe_squares
```

6.10 Reaching Knight Jump

The cheeseboard game is typically played in two dimensions, but it can be extended to k dimensions. In this game, a knight is represented by the head and neck of a horse. In a two-dimensional chessboard, a knight can move to any of its eight neighboring squares, while in a k-dimensional chessboard, it can move to squares in other dimensions as well. The objective of this challenge is to determine whether a knight can reach a given position on the chessboard, given its starting and ending coordinates. The coordinates are represented by tuples of negative or non-negative integers. For example, given a knight's initial position of (2, 1, 7) and starting and ending coordinates of (3, 5, 9) and (8, 11, 13), respectively, the challenge is to determine whether the knight can reach its destination or not.

It is subtracted each value in the start tuple from each value in the stop tuple and takes an absolute value from each result. The subtraction of |3 − 8|, |5 − 11| and |9 − 13| are (5, 6, 4), respectively. The given position of the knight must be exactly equal to the result of subtraction. Hence, as (2, 1, 7) is not equal to (5, 6, 4), (2, 1, 7) is not reachable from the given start and stop positions. It is should be noted that if c is the subtraction result, c must be sorted in descending order. Write a function that as input, takes knight, start and stop coordinates, and returns True if the knight is reachable from start and stop coordinates, else returns False. For some inputs, the expected outputs are illustrated in Table 6.10.

Algorithm

1. It takes in three arguments: *knight*, *start*, and *end*. It returns True if a knight can move from the start position to the end position on a chessboard, and False otherwise.
2. It initializes an empty list called *list_numbers*.
3. It loops through each element of the start position and computes the absolute difference between the corresponding element in the start and end position.
4. It then appends this difference to the *list_numbers*.
5. It sorts the *list_numbers* in descending order using the *bubble_sort*.
6. It loops through each element of the *knight* list. If any element of *knight* is not in *list_numbers*, it means that the knight cannot reach the target position. In this case, the function returns False.
7. If all the elements of knight are present in *list_numbers*, it means that the knight can reach the target position. In this case, the function returns True.

The Python code for determining whether knight chess is reachable from the given coordinates of the start and stop is depicted in Code 6.10.

Code 6.10 Python code for determining whether a knight chess is reachable from the given coordinates of start and stop

```
'''
perform bubble sort on an input array
and return a reversed sorted array
'''
def bubble_sort(arr):
    n = len(arr)
    for i in range(n - 1):
        for j in range(0, n - i - 1):
            if arr[j] < arr[j + 1]:
                '''
                Swap the two elements if
                they are in the wrong order
                '''
                arr[j], arr[j + 1] = arr[j + 1], arr[j]
    return arr
'''
To determine whether a knight can reach
a given position on a k-dimensional chessboard
'''
def reaching_knight_jump(knight, start, end):
    '''
    # Initialize an empty list to hold the
    differences between start and end positions
    '''
    list_numbers = []
```

```
26      for i in range(len(start)):
27          '''
28          # Compute the absolute difference
29          between each element of start and end
30          and append to list_numbers
31          '''
32          list_numbers.append(abs(start[i] - end[i]))
33      # Sort the list of differences in descending order
34      list_numbers = bubble_sort(list_numbers)
35      for num in knight:
36          '''
37          If any element of 'knight' is not
38          in 'list_numbers',
39          the knight cannot reach
40          the target position
41          '''
42          if num not in list_numbers:
43              return False
44      '''
45      Otherwise, return 'True', indicating that
46      the knight can reach the target position
47      '''
48      return True
```

6.11 Capturing Maximum Checkers

On an $n - by - n$ chessboard with black and white squares, there exists a variant known as the checker game. This game includes a king piece with coordinates (x, y) and all positions of the opponent's pawns. Each player can move one of their pieces one square in their turn, but the pieces must always remain on the white squares and be moved forward. If two pieces of different colors are diagonally adjacent, the player can jump over the opponent's piece and capture it, provided that the back of the opponent's piece is empty. If a player has the opportunity to capture an opponent's piece, they must do so, and if there are multiple pieces that can be captured in a sequence, the player must capture all of them in one move. The player who loses all their pieces loses the game. In this challenge, the objective is to capture the maximum number of opponent's pieces in one move with a king piece located at coordinates (x, y). The king can only move in the directions $(1, 1), (1, -1), (-1, 1)$, and $(-1, -1)$, and can capture an opponent's piece if it is located one square away and there is an empty space behind it. Write a function that as input, takes the dimension of the chessboard, the coordinate of a king, and the coordinate of the pieces for the

Table 6.11 The expected outputs for certain inputs for counting the maximum checkers

n, x, y, pieces	Expected output
5,1,3, set([(2,4), (2,3),(2,2),(1,2),(3,3),(1,3)])	1
8,0,4,set([(2,1), (3,1), (1,3),(4,1), (1,4)])	2
10,4,7,set([(2,1), (9,1),(2,1), (6,5)])	0

opponent, and returns the number of the most captured pieces in one move. For some inputs, the expected outputs are illustrated in Table 6.11.

Algorithm

The algorithm takes as input the size of the chessboard 'n', the starting coordinates of the king 'x' and 'y', the positions of the opponent's pieces 'pieces', and the number of pieces already removed 'removed'. A BFS based algorithm is used to explore all possible moves from the current state of the board. It initializes a set called 'visited' to keep track of the states already visited during the breadth-first search and a *queue* containing the starting state. It also initializes the current maximum to the number of checkers already removed. The algorithm then enters a while loop and dequeues the next state from the *queue*. It updates the current maximum to the maximum of the current maximum and the number of checkers removed so far in this state. For each possible diagonal jump direction, the algorithm computes the new position after the jump and the position of the jumped-over checker. If the new position is outside the board or there is already a checker at the new position or there is no checker at the jumped-over position, the algorithm skips this direction. If the algorithm finds a new state with a removed checker, it creates a new set of pieces without the jumped-over checker and checks if this state has already been visited. If not, the algorithm adds the new state to the visited set and the *queue*. The algorithm continues the breadth-first search until the *queue* is empty and returns the current maximum number of removed checkers. The Python code to capture the most pieces in one move is depicted in Code 6.11.

6.11 Capturing Maximum Checkers

Code 6.11 Python code to capture the most pieces in one move

```python
'''
This function adds an item
to the end of the queue
'''
def enqueue(queue, item):
    queue.append(item)

'''
This function removes and
returns the first item from the queue
'''
def dequeue(queue):
    return queue.pop(0)

'''
This function checks if the queue is empty
'''
def is_empty(queue):
    return len(queue) == 0

'''
This function finds the maximum number of
 checkers that can be captured
'''
'''
by a single checker at position (x, y)
on an n x n checkerboard,
'''
'''
given the positions of the other
 checkers in the set 'pieces',
'''
'''
and the number of removed
checkers so far 'removed'.
'''
def Capturing_Max_Checkers(n, x, y, pieces, removed=0):
    '''
    We use a set called 'visited'
    to keep track of the states
    '''
    visited = set()
    '''
```

```
44      '''
45      We start the search with a queue
46      containing the starting state
47      '''
48      queue = [(x, y, pieces, removed)]
49      '''
50      We initialize the current maximum to the
51      number of checkers that have
52      already been removed
53      '''
54      current_max = removed
55
56      '''
57      While the queue is not empty,
58      we continue the breadth-first search
59      '''
60      while not is_empty(queue):
61          '''
62          We dequeue the next state from the queue
63          '''
64          x, y, pieces, removed = dequeue(queue)
65          '''
66          We update the current maximum to the maximum
67          of the current maximum and the number
68          of checkers that have been removed so
69          far in this state
70          '''
71          current_max = max(current_max, removed)
72
73          # For each possible diagonal jump direction
74          for dx,dy in [(1, 1),(-1, 1),(1, -1),(-1, -1)]:
75              '''
76              We compute the new position after the jump
77              and the position of the jumped-over checker
78              '''
79              new_x, new_y = x + 2*dx, y + 2*dy
80              jump_x, jump_y = x + dx, y + dy
81
82              '''
83              If the new position is outside the board,
84              we skip this direction
85              '''
86              if not (0 <= new_x < n and 0 <= new_y < n):
87                  continue
88
                  '''
```

6.11 Capturing Maximum Checkers

```python
            '''
            If there is already a checker at the new
            position, we skip this direction
            '''
            if (new_x, new_y) in pieces:
                continue

            '''
            If there is no checker at
            the jumped-over position,
            we skip this direction
            '''
            if (jump_x, jump_y) not in pieces:
                continue

            '''
            We create a new set of pieces
            where removed the jumped-over checker
            '''
            new_pieces = pieces.copy()
            new_pieces.remove((jump_x, jump_y))

            '''
            If we have already visited this
            state before, we skip it
            '''
            if (new_x, new_y, tuple(new_pieces))\
            in visited:
                continue

            '''
            It is add the new state to the
            visited set and the queue
            '''
            visited.add((new_x, new_y, \
            tuple(new_pieces)))
            enqueue(queue,(new_x,new_y,\
            new_pieces,removed+1))

    '''
    Returning the current maximum
    number of removed checkers
    '''
    return current_max
```

6.12 Safe Rooks with Friends

Chess is a sport that is familiar to everyone, or at the very least, has been heard of. However, have you ever pondered the question of how the game would progress if only the rook pieces were present on the board? Consider an $n-by-n$ chessboard with a selection of rook pieces placed upon it, some of which belong to the same side as the friend player, and some of which belong to the opposing side. The rook pieces are permitted to move either horizontally or vertically, and the chessboard spans from square $(0, 0)$ to $(n-1, n-1)$. The objective of this challenge is to determine the number of safe squares on the board for friend rooks. For some inputs, the expected outputs are illustrated in Table 6.12.

Algorithm

It takes three inputs: an integer n representing the size of a square chessboard, a list of tuples $friend\_rooks$ representing the positions of friendly rooks on the board, and a list of tuples $enemy\_rooks$ representing the positions of enemy rooks on the board. It returns an integer representing the number of safe squares on the board. In the first step, a board of size $n * n$ is created, with all squares initialized to zero to signify that they are empty. Friend's pieces are then placed on the board represented by the value 1, while enemy pieces are represented by -1. In the next step, the safety of each square is checked by examining its surrounding squares in the left, right, up, and down directions. If a square is safe from all these directions, meaning it does not contain an enemy piece, or it contains a friend's piece, or it is empty, then it is considered a safe square. The counter variable $Safe_squares$ is incremented by one for each safe square encountered. Finally, in the last step, the function returns $Safe_squares$, which represents the number of safe squares on the board for the friend's pieces.

The Python code to return the number of safe squares for friend rooks is depicted in Code 6.12.

Table 6.12 The expected outputs for certain inputs for determining the number of safe squares on the board for friend rooks

n,friends,enemies	Expected output
22,[(11, 7), (2, 4), (15, 7)], [(10, 20), (18, 12)]	397
9, [(5,5)], [(2,5), (1,3)]	52
4, [(2,2)], [(2,1)]	10
7, [(2,2)], [(2,1),[6,4],[6,3]]	22

6.12 Safe Rooks with Friends

Code 6.12 Python code to return the number of safe squares for friend rooks

```python
 1  def Safe_rooks_with_friends(n, friend_rooks, enemy_rooks):
 2      # Initialize the chessboard
 3      chess = [[0] * n for _ in range(n)]
 4      # Place rooks on the board
 5      for x, y in friend_rooks + enemy_rooks:
 6          value = 1 if (x, y) in friend_rooks else -1
 7          chess[x][y] = value
 8
 9      # Count safe squares
10      safe_squares = 0
11      for i in range(n):
12          for j in range(n):
13              if chess[i][j] != 0:
14                  continue # Skip occupied squares
15
16              # Check if the square is safe
17              is_safe = True
18              for k in range(j - 1, -1, -1): # Check left
19                  if chess[i][k] == -1:
20                      is_safe = False
21                      break
22                  elif chess[i][k] == 1:
23                      break
24              for k in range(j + 1, n): # Check right
25                  if chess[i][k] == -1:
26                      is_safe = False
27                      break
28                  elif chess[i][k] == 1:
29                      break
30              for k in range(i + 1, n): # Check down
31                  if chess[k][j] == -1:
32                      is_safe = False
33                      break
34                  elif chess[k][j] == 1:
35                      break
36              for k in range(i - 1, -1, -1): # Check up
37                  if chess[k][j] == -1:
38                      is_safe = False
39                      break
40                  elif chess[k][j] == 1:
41                      break
42
43              if is_safe:
```

```
44                        # The square is safe
45                        safe_squares += 1
46            return safe_squares
```

6.13 Highest Score in Crag Score

Crag score refers to a type of dice game where, during their turn, the player rolls three dice simultaneously to obtain a score. The objective of this challenge is to determine and return the highest possible score from the three given dice. The rules specified for this challenge are illustrated in the below table. If the obtained score does not correspond to any category in below table, then the sum of the most frequently occurring number should be returned as the score. Write a function that as input, takes the outcome of the first three rolls, and returns the highest possible score.

Category	Description	Score	Instance
Crag	Any pair its sum is 13	50	
Thirteen	Any combination its sum is 13	26	
Three-Of-A-Kind	Any triple with the same face	25	
Low Straight	1-2-3	20	
High Straight	4-5-6	20	
Odd Straight	1-3-5	20	
Even Straight	2-4-6	20	

For some inputs the expected outputs are illustrated in Table 6.13.

Table 6.13 The expected outputs for certain inputs for determining the highest score in crag game

Dice	Expected output
[3,2,1]	20
[4,4,4]	25
[6,6,6]	25
[1,3,6]	6

6.13 Highest Score in Crag Score

Algorithm

It first creates two arrays, index and *Sum_of_each_Number*, both with a size of 6. The index array is used to keep track of the number of times each possible number appears in the dice list, while the *Sum_of_each_Number* array is used to store the sum of each number in the dice list given based on its index. It then loops through the first three items in the dice list, updating the index and *Sum_of_each_Number* arrays accordingly. Next, it uses a series of if-else statements to check for various combinations of numbers in the dice list that correspond to specific scoring rules. If a match is found, the score is updated accordingly. If no matching condition is found, it returns the sum of the most frequent numbers in the dice list. The Python code to return the highest possible score in the crag dice game is depicted in Code 6.13.

Code 6.13 Python code to return the highest possible score in crag dice game

```
def Highest_Score_in_Crag_Score(dice):
    '''
    Defining an array with size 6,
    as each dice is consisted of six numbers.
    '''
    index = [0]*6
    # Sum of each number in its index
    Sum_of_each_Number = [0]*6
    Score = 0
    for item in range(3):
        index[dice[item]-1] = index[dice[item]-1]+1
        
        Sum_of_each_Number[dice[item] -1] = \
        Sum_of_each_Number[dice[item]-1]+dice[item]
    '''
    According to the defined rules, the following
    if and else are defined.
    '''
    # Crag
    if (index[0] == 1 and index[5] == 2)\
    or (index[2] == 1
    and index[4] == 2) or \
    (index[4] == 1 and index[3] == 2):
        Score = 50
    # Three-Of-A-Kind
    elif (index[0] == 3 \
    or index[1] == 3 or index[2] == 3 or
    index[3] == 3 or index[4] == 3\
    or index[5] == 3):
        Score = 25
    # Thirteen
```

```
32        elif ((index[2] == 1 and index[3] == 1 \
33        and index[5] == 1) or
34        (index[1] == 1 and index[4] == 1\
35        and index[5] == 1)):
36            Score = 26
37        #246
38        elif index[1] == 1 and\
39        index[3] == 1 and index[5] == 1:
40            Score = 20
41
42        #456
43        elif index[3] == 1 and \
44        index[4] == 1 and index[5] == 1:
45            Score = 20
46        #123
47        elif index[0] == 1 and \
48        index[1] == 1 and index[2] == 1:
49            Score = 20
50        #135
51        elif index[0] == 1 and \
52        index[2] == 1 and index[4] == 1:
53            Score = 20
54        else:
55         for i in range(6):
56            '''
57            If score is less than its number
58            in a position, so Score is equal
59            to the number, because it is
60            greater than Score
61            '''
62            if Score < Sum_of_each_Number[i]:
63                Score = Sum_of_each_Number[i]
64    return Score
```

6.14 Optimal Crag Score with Multiple Rolls

This challenge is similar to the previous one, such that instead of a single roll, multiple rolls are given, where each roll is the outcome of three dice, and there is a constraint that a category can not be visited more than once. Write a function that as input, takes the outcome of the multiple rolls, and returns the highest possible score, otherwise, returns zero. For some inputs, the expected outputs are illustrated in Table 6.14.

6.14 Optimal Crag Score with Multiple Rolls

Table 6.14 The expected outputs for certain inputs for obtaining the highest score in optimal crag score with multiple rolls

Rolls	Expected output
[(3, 3,3), (2, 5, 5), (1, 5, 6), (2,3,3)]	47
[(1,4,6), (2, 3, 5), (1, 5, 1)]	14
[(1,1,1), (4,3,6)]	51

Algorithm

The algorithm is the same as the crag dice game but there are significant differences. It take multiple rolls as input. The first step in this algorithm is to generate all possible combinations of the given rolls. Next, it iterates over each combination and checks each 'dice' in the combination against a set of rules to determine its score. The rules are provided in the previous section. The *category_is_used* list is used to keep track of which rules have already been applied to the current combination. If a rule has already been applied, it cannot be used again for the same combination. The score for each 'dice' is determined based on which rule it satisfies. If a dice does not satisfy any of the rules, then its score is the sum of its values (the most repeated one). The total score for the combination is the sum of the scores. Finally, it returns the maximum score across all combinations. The Python code to return the highest possible score in the optimal crag dice game is depicted in Code 6.14.

Code 6.14 Python code to return the highest possible score in optimal crag dice game

```
'''
This function generates all
 possible combinations of n rolls
'''
def generate_combinations(rolls, n):
    if n == 0:
        return [[]]
    else:
        combinations = []
        for i in range(len(rolls)):
            sub_combinations =\
            generate_combinations(rolls, n-1)
            for combination in sub_combinations:
                if i not in combination:
                    combinations.append([i]+combination)
        return combinations

'''
This function calculates the optimal
crag score given a list of rolls
'''
```

```python
22  def Optimal_Crag_Score_with_Multiple_Rolls(rolls):
23      posibilities = generate_combinations(rolls, len(rolls))
24      unique_posibilities = []
25      for roll in posibilities:
26          if set(roll) == {i for i in range(len(rolls))}:
27              unique_posibilities.append(roll)
28      result = []
29
30      category_is_used = [False] * 13
31      for posibility in unique_posibilities:
32          category_is_used = [False] * 13
33          temp_roll = [rolls[i] for i in posibility]
34          points = []
35          for dice in temp_roll:
36              # as it is the permutation of six numbers
37              index = [0] * 6
38              Sum_of_each_Number = [0] * 6
39              Desired_Score = 0
40              for item in range(3):
41                  index[dice[item] - 1]=\
42                  dice.count(dice[item])
43                  Sum_of_each_Number[dice[item] - 1] = \
44                  dice.count(dice[item]) * dice[item]
45              '''
46              # Check if the dice qualifies for any of the
47              categories and assign the appropriate score
48              '''
49              #Crag
50              if ((index[0] == 1 and index[5] == 2) or
51                  (index[2] == 1 and index[4] == 2) or (
52                  index[4] == 1 and index[3] == 2)) \
53                  and not category_is_used[0]:
54                  Desired_Score = 50
55                  category_is_used[0] = True
56              #Three-Of-A-Kind
57              elif ((index[2] == 1 and index[3] == 1
58                  and index[5] == 1) or (
59                  index[1] == 1 and index[4] == 1
60                  and index[5] == 1))\
61                  and not category_is_used[1]:
62                  Desired_Score = 26
63                  category_is_used[1] = True
64              # Thirteen
65              elif (index[0] == 3 or index[1] == 3 or
66                  index[2] == 3 or index[3] == 3 or
```

6.14 Optimal Crag Score with Multiple Rolls

```
67                    index[4] == 3 or index[5] == 3) \
68                    and not category_is_used[2]:
69                        Desired_Score = 25
70                        category_is_used[2] = True
71                    #246
72                    elif (index[1] == 1 and index[3] == 1
73                    and index[5] == 1) \
74                    and not category_is_used[3]:
75                        Desired_Score = 20
76                        category_is_used[3] = True
77                    #456
78                    elif (index[3] == 1 and index[4] == 1
79                    and index[5] == 1) and not \
80                        category_is_used[4]:
81                        Desired_Score = 20
82                        category_is_used[4] = True
83                    #123
84                    elif (index[0] == 1 and index[1] == 1
85                    and index[2] == 1) and not \
86                        category_is_used[5]:
87                        Desired_Score = 20
88                        category_is_used[5] = True
89                    #135
90                    elif (index[0] == 1 and index[2] == 1
91                    and index[4] == 1) and not \
92                        category_is_used[6]:
93                        Desired_Score = 20
94                        category_is_used[6] = True
95
96                    else:
97                        temp_index = 0
98                        for i in range(6):
99                            if Desired_Score \
100                           < Sum_of_each_Number[i]\
101                           and not category_is_used[6+i]:
102                               temp_index = i
103                               Desired_Score =\
104                               Sum_of_each_Number[i]
105                        category_is_used[6+temp_index]=True
106                   points.append(Desired_Score)
107              result.append(sum(points))
108        try:
109            return max(result)
110        except:
111            return 0
```

Chapter 7
Count

This chapter talks about 8 count-based problems. These challenges are explained with some examples and then programmed in Python. The problems are listed as follows:

1. Counting the number of carries when addition two given numbers
2. Counting the number of animals that are growling animals
3. Counting the number of consecutive summers for a polite number
4. Counting the occurrence of each digit and read it aloud
5. Counting the number of maximal layers on two-dimensional plane
6. Troika counter in positive integers
7. Counting the number of disks that are intersected.

7.1 Counting Number of Carries

When adding two or more digits, a carry occurs if the result of the addition exceeds the largest number that the system can represent. This challenge requires counting the number of carries that occur when two numbers are added together. To accomplish this, you need to write a function that takes two positive integers, 'a' and 'b', as input and returns the number of carries. For some inputs, the expected outputs are illustrated in Table 7.1.

Algorithm
It is used two algorithm for counting the number of carries. The first calls a recursive algorithm ($get\_digit\_sum\_recursive$) inside it. The recursive algorithm computes the sum of given digits of a, b, and $a + b$. In fact, it repeatedly removes the last digit of the number using integer division ($n//10$) and adds it to the sum until there are no more digits left. Th recursive algorithm has the following steps.

Table 7.1 The expected outputs for certain inputs for counting the number of carries

a, b	Expected output
434, 289	2
2, 8	1
3, 3	50
8856, 5936	3

1. 1. Start with a positive integer 'number'.
2. If 'number' is less than 10, return 'number'.
3. Otherwise, compute the sum of the last digit of 'number' and the sum of the remaining digits of 'number' by taking the modulus of 'number' with 10 and dividing 'number' by 10, respectively.
4. Recursively call do the above steps.
5. Return the result of the recursive call.

The first algorithm has the following steps.

1. It takes two positive integers 'num1' and 'num2'.
2. Compute the sum of digits of 'num1' using the 'get_digit_sum_recursive' and store it in a variable 'digit_sum_num1'.
3. Compute the sum of digits of 'num2' using the 'get_digit_sum_recursive' and store it in a variable 'digit_sum_num2'.
4. Compute the sum of digits of 'num1 + num2' using the 'get_digit_sum_recursive' function and store it in a variable 'digit_sum_num1_num2'.
5. Compute the total number of carries as '(digit_sum_num1 + digit_sum_num2 - digit_sum_num1_num2) // 9'.
6. Return the total number of carries.

The Python code for returning and counting the number of carries is depicted in Code 7.1.

Code 7.1 Python code for returning and counting the number of carries

```
1   '''
2   computing the sum of digits of a
3   given number using recursion
4   '''
5   def get_digit_sum_recursive(number):
6       # If the number has only one digit
7       if number < 10:
8           # Return that digit
9           return number
10      else:
11          '''
12          Otherwise, add the last digit to
13          the sum of the remaining digits
14          '''
```

7.2 Counting Growl of Animals

```
15            return number % 10 + \
16                get_digit_sum_recursive(number // 10)
17
18    '''
19    Counting the number of carries that
20    occur when adding two numbers
21    '''
22    def count_carries(num1, num2):
23        # Compute the sum of digits of num1
24        digit_sum_num1 = get_digit_sum_recursive(num1)
25        # Compute the sum of digits of num2
26        digit_sum_num2 = get_digit_sum_recursive(num2)
27        # Compute the sum of digits of num1+num2
28        digit_sum_num1_num2 = get_digit_sum_recursive
29                              (num1 + num2)
30        '''
31        The total number of carries is equal to this
32        '''
33        return (digit_sum_num1 + digit_sum_num2\
34                - digit_sum_num1_num2) // 9
```

7.2 Counting Growl of Animals

There are animals that see 'cat' and 'dog' facing left, and see 'tac' and 'god' facing right. The animals are to be growling if they see strictly more dogs than cats on the right or left. For example, if cd=['god', 'cat', 'cat', 'tac', 'tac', 'dog', 'cat', 'god'] be the given cats and dogs, count the number of animals is growling. In cd, it colorized 'cat' and 'tac' that there are more dogs than cats. Write a function that as input takes a list of animals and returns the number of animals that are growling. For some inputs, the expected outputs are illustrated in Table 7.2.

Table 7.2 The expected outputs for certain inputs for counting the growl of animals

Animals	Expected output
['tac', 'tac', 'tac','god', 'tac', 'dog', 'dog']	4
['tac', 'dog', 'dog','god', 'tac','dog', 'tac']	2
['tac', 'dog', 'dog','god', 'tac']	1
['tac', 'tac', 'dog','cat', 'tac']	0

Algorithm

It iterates over each animal in the 'animals' list given and checks if it is a cat, dog, tac, or god. The algorithm has the following steps.

1. It takes list *animals* as input.
2. Initialize a variable *num_growlers* to keep track of the number of growlers seen so far.
3. Initialize a variable *current_animal_index* to zero to keep track of the current animal being checked.
4. While there are still animals left to check (i.e., *current_animal_index* is less than of length of animals:
   a. If the current animal is a cat or dog, then count the number of cats and dogs seen before it in the list. If there are more dogs than cats, then increment *num_growlers*.
   b. If the current animal is a tac or god, then count the number of cats and dogs seen after it in the list. If there are more dogs than cats, then increment *num_growlers*.
   c. Move on to the next animal in the list.
5. Return the total number of growlers found.

The Python code for returning and counting the number of animals that are growling is depicted in Code 7.2.

Code 7.2 Python code for returning and counting the number of animals are growling

```
 1  def Count_the_Growl_of_Animals(animals):
 2      # initialize the number of growlers to zero
 3      num_growlers = 0
 4      # initialize the index of the current animal to zero
 5      current_animal_index = 0
 6      # while there are still animals left to check
 7      while current_animal_index < len(animals):
 8          # if the current animal is a cat or a dog
 9          if animals[current_animal_index] in ["cat","dog"]:
10              '''
11              initialize the number of cats seen
12              so far to zero
13              '''
14              num_cats = 0
15              '''
16              initialize the number of dogs seen
17              so far to zero
18              '''
19              num_dogs = 0
20              '''
21              start checking animals
```

7.2 Counting Growl of Animals

```
            before the current one
            '''
            previous_animal_index=\
            current_animal_index - 1

            '''
            count the number of cats and dogs
            seen before the current animal
            '''
            while previous_animal_index >= 0:
                if animals[previous_animal_index]=="cat" \
                or animals[previous_animal_index]=="tac":
                    num_cats += 1
                elif animals[previous_animal_index]=="dog" \
                or animals[previous_animal_index]=="god":
                    num_dogs += 1
                previous_animal_index -= 1
            '''
            if there are more dogs than cats seen before
            the current animal,
            add it to the count of growlers
            '''
            if num_dogs > num_cats:
                num_growlers += 1
        # if the current animal is a tac or a god
        elif animals[current_animal_index] in ["tac","god"]:
            '''
            initialize the number of cats seen
            so far to zero
            '''
            num_cats = 0
            '''
            initialize the number of dogs seen
            so far to zero
            '''
            num_dogs = 0
            '''
            start checking animals
            after the current one
            '''
            next_animal_index=\
            current_animal_index + 1

            '''
            count the number of cats and dogs
```

```
67                 seen after the current animal
68                 '''
69                 while next_animal_index < len(animals):
70                     if animals[next_animal_index]=="cat" \
71                     or animals[next_animal_index]=="tac":
72                         num_cats += 1
73                     elif animals[next_animal_index]=="dog"\
74                     or animals[next_animal_index]=="god":
75                         num_dogs += 1
76                     next_animal_index += 1
77
78                 '''
79                 if there are more dogs than cats
80                 seen after the current animal, add
81                 it to the count of growlers
82                 '''
83                 if num_dogs > num_cats:
84                     num_growlers += 1
85             # move on to the next animal
86             current_animal_index += 1
87         # return the total number of growlers found
88         return num_growlers
```

7.3 Counting Consecutive Summers for a Polite Number

In the domain of number theory, a positive integer that can be expressed as the sum of two or more consecutive numbers is referred to as a polite number. On the other hand, a positive integer that cannot be expressed in this manner is categorized as an impolite number. It is important to note that both impolite and polite numbers are considered natural numbers. This challenge is aimed to count the number of ways a number can be expressed as consecutive positive integers. For example, the consecutive positive integers for 42 are (a) $3+4+5+6+7+8+9$, (b) $9+10+11+12$, (c) $13+14+15$, and (d) 42, or for 3, the consecutive positive integers are $2+1$ and 3. Write a function that takes a positive integer n and returns the number of ways a number can be expressed as consecutive positive integers. For some inputs, the expected outputs are illustrated in Table 7.3.

7.4 Counting Occurrence of Each Digit

Table 7.3 The expected outputs for certain inputs for counting the consecutive summers

n	Expected output
3	2
96000	8
2	1
42	4

Algorithm

The algorithm will count the number of odd divisors of 'n', which is equivalent to the number of ways in which 'n' can be expressed as a sum of consecutive positive integers.

1. Take a positive integer 'n' as input.
2. Initialize a variable called 'NumberofPolites' to 0.
3. Use a loop to iterate over the range '1' to 'n+1'.
4. For each integer 'i' in the loop:
   a. Check if 'n' is divisible by 'i' and 'i' is odd.
   b. If both conditions are satisfied, increment the 'NumberofPolites' variable by 1.
5. Return 'NumberofPolites' as the output.

The Python code for returning and counting the number of consecutive summers is depicted in Code 7.3.

Code 7.3 Python code for returning and counting the number of consecutive summers

```
1  def count_consecutive_summers(n):
2      NumberofPolites = 0
3      for i in range(1, n + 1):
4          if n % i == 0 and i % 2 == 1:
5              NumberofPolites += 1
6      return NumberofPolites
```

The for loop iterates over all integers from 1 to n+1. The if statement checks if the current integer 'i' is a divisor of n and whether it is odd or not. If both conditions are satisfied, then the counter variable *NumberofPolites* is incremented by 1. Therefore, at the end of the loop, the value stored in *NumberofPolites* represents the total count of odd divisors of *n*.

7.4 Counting Occurrence of Each Digit

There is a string of digits that contains only digits from '0123456789'. It is aimed to read aloud each digit that is repeated several times. For example, if '222274444499966' is the given string, the output is '4217543926'. From left to right, it means two times 4, one time 7, five times 4, three times 9, and two times 6 are repeated. Write a function that as input takes a string of digits and reads it aloud as instructed. For some inputs, the expected outputs are illustrated in Table 7.4.

Table 7.4 The expected outputs for certain inputs for counting the occurrence of each digit

Digits	Expected output
'7779981'	'37291811'
'1333334'	'115314'
'2115131114'	'12211511133114'
'37291811'	'13171219111821'

Algorithm

It takes a list of digits as input and iterates through it, maintaining a count of consecutive identical digits using a variable named count. When a different digit is encountered, it appends the count and digit to a list named 'orders' and resets the count to 1. Finally, it returns a string representation of the elements in the 'orders' list.

The Python code to read aloud each digit is depicted in Code 7.4.

Code 7.4 Python code to read aloud each digit languagelanguage

```
1   def Count_occurrence_of_each_digit(digits):
2       orders = []
3       count = 1
4       previous_digit = digits[0]
5
6       '''
7       Iterate through the digits and
8       count the occurrence of each digit
9       '''
10      for current_digit in digits[1:]:
11          if current_digit == previous_digit:
12              count += 1
13          else:
14              '''
15              Append the count of the previous digit
16              and the digit itself to the result
17              '''
18              orders.append(str(count))
19              orders.append(previous_digit)
20              count = 1
21              previous_digit = current_digit
22
23      '''
24      Append the count of the last digit
25      and the digit itself to the result
26      '''
27      orders.append(str(count))
28      orders.append(previous_digit)
29      return ''.join(orders)
```

7.5 Counting Maximal Layers

Table 7.5 The expected outputs for certain inputs to count maximal layers

Points	Expected output
[(7,1), (6, 9)]	1
[(11,4), (1, 1), (11,1)]	2
[(1, 5), (3, 10), (2, 1), (9, 2)]	2
[(796,1024), (11, 13), (19,221), (700,3)]	3

7.5 Counting Maximal Layers

On the two-dimensional plane, a point (x_1, y_1) dominates (x_2, y_2), if $x_1 > x_2$ and $y_1 > y_2$. If a point is not dominated by all points, the point is considered a maximal one, where on the two-dimensional plane there can be any number of points that form a maximal layer each time. This challenge is aimed to count the number of maximal layers. For example, if $points = [(1, 5), (3, 10), (2, 1), (9, 2)]$ be the given input, count the number of maximal layers. From left to right the input is traversed, $(1, 5)$ is not a maximal one, because it is dominated by $(3, 10)$, 3 is greater than 1 and 10 is greater than 5. The next considered point from left to right is $(3, 10)$, it is a maximal one, because for $(1, 5)$ and $(3, 10)$, 1 is not greater than 3, and 5 is not grater than 10, and for $(2, 1)$ and $(3, 10)$, 2 is not greater than 3, and 1 is not grater than 10, and for $(9, 2)$ and $(3, 10)$, 9 is greater than 3, but 2 is not grater than 10. The next considered one is $(2, 1)$ and is not a maximal one, because it is dominated by $(3, 10)$, and $(9, 2)$ is a maximal one, because alike $(3, 10)$ it is not dominated by any point. Therefore, for $points = [(1, 5), (3, 10), (2, 1), (9, 2)]$ there are two maximal layers that one is $(10, 3)$ and another is $(9, 2)$. Write a function that as input takes a list of points and returns the number of maximal layers. For some inputs, the expected outputs are illustrated in Table 7.5.

Algorithm

Let *points* be the given input, *layerCounter* be to enumerate the maximal layers. It loops through the points in the list and checks if each point is maximal (i.e., if there are no other points with greater x and y coordinates). If a point is maximal, it is added to a list of maximal points. Once all maximal points have been identified, the function increments the *layerCounter* and removes the maximal points from the original list. The loop continues until there are no more points left in the list, and the final layer count is returned. The Python code to count the number of maximal layers is depicted in Code 7.5.

Code 7.5 Python code to count the the number of maximal layers

```
1  def count_maximal_layers_in_points(points_list):
2      # Initialize layer counter to 0
3      layer_count = 0
4
5      # Loop until points_list becomes empty
```

```
 6      while points_list:
 7          # Find all maximal points in the list
 8          maximal_points = []
 9          for point in points_list:
10              is_maximal = True
11              for other_point in points_list:
12                  if other_point[0] > point[0] \
13                     and other_point[1] > point[1]:
14                      is_maximal = False
15                      break
16              if is_maximal:
17                  maximal_points.append(point)
18
19          '''
20          Increment layer_count by 1 if
21          there are any maximal points
22          '''
23          if maximal_points:
24              layer_count += 1
25
26          '''
27          Remove all maximal points
28          from the original list
29          '''
30          for point in maximal_points:
31              points_list.remove(point)
32
33      # Return the number of maximal layers
34      return layer_count
```

7.6 Counting Dominator Numbers

An element is considered a dominator in a list of items if the numbers to its right are smaller than the element. The task in this challenge is to find the number of dominators in a given list. According to this definition, the last element of the list is always a dominator. For example, consider the given list $items =$ $[-492, 124, 113, -38, -28, -15]$. We can see that element 124 is a dominator because it is larger than 113, -38, -28, and -15. Element 113 is also a dominator because it is larger than -38, -28, and -15. Finally, -15 is the last element of the list, and so it is also a dominator. Therefore, the number of dominators in this list is three. If there is an empty list as an input, a zero number must be returned. For some inputs, the expected outputs are illustrated in Table 7.6.

7.6 Counting Dominator Numbers

Table 7.6 The expected outputs for certain inputs for counting dominator numbers

items	Expected output
[−492, 124, 113, −38, −28, −15]	3
[77, 1, 2, −4, 13, 7, 9, 1]	4
[13, 7, 9, 1, 77, 1, 2, −4]	3
[77, 1, 2, −4, 13, 77, 7, 9, 1]	3

Algorithm

1. It takes a list of integer numbers.
2. Check if the input list is empty. If it is, return 0.
3. Initialize a variable max_num to negative infinity and a variable $dominator\_count$ to 0.
4. Iterate over the reversed list of items.
5. For each number in the list, check if it is greater than max_num.
6. If it is, update max_num to the current number and increment $dominator\_count$.
7. After iterating over the entire list, return $dominator\_count$.

The Python code to count the number of dominator numbers is depicted in Code 7.6.

Code 7.6 Python code to count the the number of dominator numbers

```
#Reverse the input
def reverse_list(lst):
    '''
    create an empty list to
    store the reversed elements
    '''
    reversed_lst = []
    '''iterate over the indices of
    the list in reverse order
    '''
    for i in range(len(lst)-1, -1, -1):
        reversed_lst.append(lst[i])
    # return the reversed list
    return reversed_lst
"""
Counts the number of dominators
in a list of integers.
"""
def Counting_Dominator_Numbers(items):

    # if the list is empty, return 0
    if not items:
        return 0
```

```
24
25      '''
26      In Python, -inf represents negative infinity.
27      '''
28      max_num = float('-inf')
29      # Initialize the dominator count to 0
30      dominator_count = 0
31      # iterate over the reversed list of items
32      for num in reverse_list(items):
33          '''
34          if the current number is greater than
35          the maximum number seen so far
36          '''
37          if num > max_num:
38              # update the maximum number
39              max_num = num
40              # increment the dominator count
41              dominator_count += 1
42      # return the dominator count
43      return dominator_count
```

7.7 Counting Troikas from Integer Numbers

Let *items* be a list of numbers such that positions $i < j < k$ in *items* make a troika if $items[i] == items[j] == items[k]$, and $j - i == k - j$. For example, if *items* $= [42, 9, 42, 42, 42, 103]$ is the given input, count the number of troikas. In the 0th position, 2th position, and 4th position, there are 42, 42, and 42, respectively, where $2 - 0 = 4 - 2$, so $2 = 2$. In the previous step, one Troika is found, and another Troika is in the 2th position, 3th position, and 4th position, because $3 - 2 = 4 - 3$, so $1 = 1$. Therefore two Troikas are found. Write a function that as input takes a list of positive and negative integers and returns the number of troikas. The expected outputs are illustrated in Table 7.7 for some inputs.

Table 7.7 The expected outputs for certain inputs for counting troikas

Items	Expected output
[42, 9, 42, 42, 42, 103]	2
[−8, −8, 109, −8, −8, −8, −36]	2
[21, −41, 21, 76, 21, −71, 21, 17, 17, 17, 17, 47]	4
[77, 1, 2, −4, 13, 77, 7, 9, 1]	0

Algorithm

Let $TroikaCounter$ be to enumerate the number of troikas, and $items$ be the given numbers. To satisfy condition $j - i == k - j$, it is utilized condition $items[i] == items[j] == items[2*j - i]$, and for each i in range 0 to $|items|$, and for each j in range $i + 1$ to $|items|$, if condition $items[i] == items[j] == items[2*j - i]$ is satisfied ($2*j - i$ determines 3th position), the $TroikaCounter$ is incremented by one. Now, $TroikaCounter$ is the number of troikas. The Python code to count the number of troikas is depicted in Code 7.7.

Code 7.7 Python code to count the the number of troikas

```
1  def TroikaCounter(items):
2      TroikaCounter = 0
3      for i in range(len(items)):
4          for j in range(i + 1, len(items)):
5              k = 2*j - i
6              if k > (len(items)-1):
7                  pass
8              else:
9                  if items[i] == items[j] == items[k]:
10                     TroikaCounter += 1
11     return TroikaCounter
```

7.8 Counting Intersected Disks

On the two dimensional plane, there are disks as $(x1, y1, r1)$ and $(x2, y2, r2)$, where (x, y) is the center point and r is the radius of the disks such that if Pythagorean inequality $(x2 - x1)**2 + (y2 - y1)**2 <= (r1 + r2)**2$ satisfies, two disks $(x1, y1, r1)$ and $(x2, y2, r2)$ have intersection. The power in Python is shown with $**$. It should be noted that the formula just works for the integers. In this challenge, the coordinates of the center and the radius of the disks in a list are given, and the task is to write a function that returns the pair of disks that intersect at least at one point. For some inputs, the expected outputs are illustrated in Table 7.8.

Table 7.8 The expected outputs for certain inputs for counting intersected disks

Disks	Expected output
[(1, 1, 9), (1, 0, 9)]	1
[(4,7,11), (11, 2, 8), (1, 1, 1)]	2
[(1, 1, 3), (6, 0, 3), (6, 5, 3), (0, 6, 3)]	3
[(34, 9, 67)]	0

Algorithm

To solve this challenge the sweep line algorithm is utilized. The sweep line algorithm is a popular technique used in computational geometry to solve various problems related to geometric objects' arrangement and intersection. By sweeping a line or plane across a set of objects, it enables the efficient detection of intersections, overlaps, and other geometric properties. The number of comparisons of the circles is $\frac{n \times n - 1}{2}$, where it compares all circles and consequently increases the number of comparisons and the execution time. To reduce the number of comparisons especially when the two circles have a far position, the sweep line algorithm is used, it is utilized to compute the active space. In the active space, the intersection of the circles is more frequent. Each disk enters the active space when the vertical line than to the x-axis enters space $(x - r)$ and exits from the active space when vertical line leave $(r + x)$. The Python code to count the number of overlapping disks is depicted in Code 7.8

Code 7.8 Python code for counting the overlapping disks

```
 1  def Counting_Intersected_Disks(disks):
 2      # Sort disks
 3      active_disks = sorted([(x − r, x + r, x, y, r)
 4                             for x, y, r in disks])
 5
 6      # Initialize count of overlapping disks to zero
 7      count = 0
 8
 9      '''
10      Iterate over pairs of disks
11      and count overlapping disks
12      '''
13      for i in range(len(active_disks) − 1):
14          # Start with the next disk after i
15          j = i + 1
16
17          # Check if j intersects with i
18          while j < len(active_disks)\
19          and active_disks[j][0] <= active_disks[i][1]:
20              '''Calculate the distance between
21              the centers of i and j
22              '''
23              dx = active_disks[i][2]− active_disks[j][2]
24              dy = active_disks[i][3]− active_disks[j][3]
25              r_sum =active_disks[i][4]+ active_disks[j][4]
26
27              # Check if i and j overlap
28              if dx ** 2 + dy ** 2 <= r_sum ** 2:
29                  count += 1
30
```

7.8 Counting Intersected Disks

```
31              # Move to the next disk
32              j += 1
33
34          # Return the count of overlapping disks
35          return count
```

Chapter 8
Miscellaneous Problems

This chapter discusses six miscellaneous problems. These challenges are elucidated with illustrative examples and subsequently implemented using the Python programming language. The problems are enumerated as follows:

1. Performing a perfect riffle to items
2. Exact change money into an array of coin denominations
3. Just keeping items whose Frequency is at most n
4. When two frogs are collided in the same square
5. Positioning in Wythoff Array
6. Interpreting Fractran program.

8.1 Riffling Items

This challenge involves a list of items with an even number of elements, and the goal is to perform a perfect riffle shuffle on the items. In a perfect riffle shuffle, the list is split into two equal halves, and the shuffle is performed by alternately taking one element from each half. There are two types of perfect riffle shuffles: an out-shuffle, where the first item of the first half is taken first, followed by the first item of the second half, and so on until all items have been shuffled; and an in-shuffle, where the first item of the second half is taken first, followed by the first item of the first half, and so on until all items have been shuffled. For example, if the given items be $items = [19, 456, 3, 4, 5, 6, 64, 11]$, riffle $items$ with out-shuffle order. Firstly, in the two same-sized halves $items$ is splitted, so $p_1 = [19, 456, 3, 4]$ and $p_2 = [5, 6, 64, 11]$, and st is defined to store the outputs. It is the order of insertion of outputs in storage, $st = [19] \rightarrow st = [19, 5] \rightarrow st = [19, 5, 456] \rightarrow st = [19, 5, 456, 6] \rightarrow st = [19, 5, 456, 6, 3] \rightarrow st = [19, 5, 456, 6, 3, 64] \rightarrow st = [19, 5, 456, 6, 3, 64, 4] \rightarrow st = [19, 5, 456, 6, 3, 64, 4, 11]$. So, $[19, 5, 456, 6, 3, 64, 4, 11]$ is as the riffled list. For the example above with in-shuffle order, the riffled list of items is $st = [5, 19, 6, 456, 64, 3, 11, 4]$. The process to reach the answer are $st = [5] \rightarrow st =$

Table 8.1 The expected outputs for certain inputs to make a perfect riffle

Items, out	Expected output
[19, 456, 3, 4, 5, 6, 64, 11], True	[19, 5, 456, 6, 3, 64, 4, 11]
[19, 456, 3, 4, 5, 6, 64, 11], False	[5, 19, 6, 456, 64, 3, 11, 4]
[0, 7, 89, 654, 5, 0, 1, 7]	[0, 5, 7, 0, 89, 1, 654, 7]

$[5, 19] \to st = [5, 19, 6] \to st = [5, 19, 6, 456] \to st = [5, 19, 6, 456, 64] \to st = [5, 19, 6, 456, 64, 3] \to st = [5, 19, 6, 456, 64, 3, 11] \to st = [5, 19, 6, 456, 64, 3, 11, 4]$. Write a function that as the input, takes a list of items and boolean variable and returns the riffled list of items as out-shuffle or in-shuffle. For some inputs, the expected outputs are illustrated in Table 8.1.

Algorithm

Let $items$ be a list of the given items, $storage$ be to store the outputs, and $outin$ be a boolean, where if $outin$ is $True$, mean out-shuffle, else in-shuffle. The first half is stored in $part1$, and the second half is stored in $part2$. The |items|[1] is divided by two, and is stored in $half$, and for w in $|halves|$ if $outin$ is equal to $True$, $part1[w]$ is stored into $storage$ and then $part2[w]$ is stored into $storage$, else $part2[w]$ is stored into $storage$ and then $part1[w]$ is stored into $storage$. Now, $storage$ is the list of riffled items.

The Python code to make a perfect riffle in the list of items is depicted in Code 8.1.

Code 8.1 Python code to make a perfect riffle

```
1  def perfect_riffle(items, outin=True):
2      '''
3      out=True: out-shuffle
4      out=False: in-shuffle
5      '''
6      # to store the outputs
7      storage=[]
8      # Splitting into two same-sized halves
9      halves=int(len(items)/2)
10     totallength=int(len(items))
11     # part1 is the first half
12     part1=items[0:halves]
13     # part2 is the second half
14     part2=items[halves:totallength]
15     for w in range(halves):
16         # Making out-shuffle
17         if outin==True:
```

---

[1] The length of items.

```
18              storage.append(part1[w])
19              storage.append(part2[w])
20          # Making in-shuffle
21          else:
22              storage.append(part2[w])
23              storage.append(part1[w])
24    return storage
```

A more efficient approach exists that utilizes a reduced amount of memory. The Python code to make a perfect riffle in the list of items is depicted in Code 8.2.

Code 8.2 Python code to make a perfect riffle
```
1  def perfect_riffle(items, outin=True):
2      '''
3      out=True: out-shuffle
4      out=False: in-shuffle
5      '''
6      # Calculate the number of cards in each half
7      n = len(items) // 2
8      # Split the deck into two halves
9      half1 = items[:n]
10     half2 = items[n:]
11     # Iterate over one of the halves
12     for i in range(n):
13         # Perform an out-shuffle
14         if outin:
15             items[2*i] = half1[i]
16             items[2*i+1] = half2[i]
17         # Perform an in-shuffle
18         else:
19             items[2*i] = half2[i]
20             items[2*i+1] = half1[i]
21     return items
```

8.2 Calculate Smaller Coins

Given a positive integer amount and an array of coin denominations in decreasing order, the objective is to exchange the money into smaller coins such that their total value is exactly equal to the amount. It is guaranteed that the last coin denomination in the array is 1. For example, if the amount is 583 and the coin denominations are [142, 73, 45, 17, 13, 7, 1], the amount 583 must be converted into [142, 142, 142, 142, 13, 1, 1]. Write a function that takes as input a positive integer amount and an array of coin denominations, and returns an array of coin denominations whose total value is exactly equal to the input amount. For some inputs, the expected outputs are illustrated in Table 8.2.

Table 8.2 The expected outputs for certain inputs for calculating smaller coins

Amount, coins	Expected output
583, [142, 73, 45, 17, 13, 7, 1]	[142, 142, 142, 142, 13, 1, 1]
26, [142, 45, 7, 1]	[7, 7, 7, 1, 1, 1, 1, 1]
174, [94, 73, 17, 13, 7, 4, 3, 1]	[94, 73, 7]
1, [142, 3, 1]	[1]

Algorithm

1. Let *amount* be the given amount and *coins* be the list of coins.
2. Initialize an empty list called 'smallercoins' to store the smaller coins that make up the amount.
3. For each coin in the 'coins' list, repeat steps 4–6:
4. Check if the remaining 'amount' is greater than or equal to the current 'coin'.
5. If the remaining 'amount' is greater than or equal to the current 'coin', subtract the 'coin' from the 'amount' and add it to the 'smallercoins' list.
6. If the remaining 'amount' is less than the current 'coin', move on to the next coin in the 'coins' list.
7. Once all coins have been checked, return the 'smallercoins' list containing the smaller coins that make up the amount.

The Python code to change the monies into the smaller coins depicted in Code 8.3.

Code 8.3 Python code to change the monies into the smaller coins

```
1  def calculate_smaller_coins(amount, coins):
2      # Create an empty list to store the smaller coins
3      smallercoins = []
4      # Loop through each coin in the list of coins
5      for coin in coins:
6          '''Keep subtracting the coin value from the
7          amount until it becomes less than the coin value
8          '''
9          while amount >= coin:
10             '''
11             Add the coin value to the
12             list of smaller coins
13             '''
14             smallercoins.append(coin)
15             # Subtract the coin value from the amount
16             amount -= coin
17     # Return the list of smaller coins
18     return smallercoins
```

Table 8.3 The expected outputs for certain inputs for getting items whose frequency is at most n

Items, n	Expected output
[4, 4, 4, 4, 4, 4], 4	[4, 4, 4, 4]
[4, 5, 1000, 1, 2, 3], 5	[4, 5, 1000, 1, 2, 3]
[0, 0, 0, 0, 2, 0, 0, 0,0], 5	[0, 0, 0, 0, 2, 0]
[1001,'python',1003,'world',1001,1001]	[1001, 'python', 1003, 'world', 1001]

8.3 Keep Frequent Items at Most n

There is a list of items with an arbitrary length, that is aimed to generate a list of items that just keep exactly the items whose frequencies are less than or equal to n, where the positions in the new list are the same as the original list. For example, consider if i = ['*Python*', 4, 99, '*R*', 4, '43', 123, 80, 99, '*Python*', 4] be the given items and $n = 2, j$ = ['*Python*', 4, 99, '*R*', 4, '43', 123, 80, 99, '*Python*'] is the new list that just keeps the items that their frequency is less than or equal with n. In list i, except for 4 that is repeated three times, all items are stored into list j, and as $n = 2$ so the last 4 is skipped, because 4 is repeated three times. Write a function that takes as input a positive integer n and a list of items, and returns a list of items whose frequency is at most n and whose order in the list is preserved. If n = 0, the function should return an empty list. For some inputs, the expected outputs are illustrated in Table 8.3.

Algorithm

It takes *items_list* and *max_frequency* as inputs and initializes an empty list *top_frequency_items* to store the items whose frequencies are less than or equal to *max_frequency*. Then, it creates an empty dictionary with keys as the items in the list and initializes their values to 0. It counts the frequency of each item in the list by iterating over the list and incrementing the value of the corresponding key in the dictionary. It then ensures that the frequency of each item is not greater than the given *max_frequency* by iterating over the dictionary and setting the value to *max_frequency* if it exceeds it, and stores the resulting dictionary in *top_frequency_dict*. Then, it iterates over *items_list* and adds each item to the *top_frequency_items* list if its frequency in *top_frequency_dict* is greater than 0 and decrements its frequency in the dictionary by 1. After traversing all items in *items_list*, *top_frequency_items* is returned as the output. The Python code generates a new list that each item is repeated with a maximum of *n* depicted in Code 8.4.

Code 8.4 Python code to generate a new list that each item is repeated with a maximum of *n*

```
1   def calculate_item_frequency(items_list, max_frequency):
2       '''
3       This function calculates the frequency of each
4       item in the given list and returns a dictionary
```

```
 5      with the item as the key and its frequency
 6      as the value. It also ensures that the
 7      frequency of each item is
 8      not greater than the given max_frequency.
 9      '''
10      frequency_dict = dict.fromkeys(items_list, 0)
11      for item in items_list:
12          frequency_dict[item] += 1
13      for key in frequency_dict:
14          if frequency_dict[key] > max_frequency:
15              frequency_dict[key] = max_frequency
16      return frequency_dict
17
18  def items_below_max_frequency(items_list, max_frequency=1):
19      '''
20      This function keeps the items with frequency
21      not greater than the given max_frequency.
22      '''
23      if max_frequency == 0:
24          return []
25      top_frequency_dict = \
26      calculate_item_frequency(items_list, max_frequency)
27      top_frequency_items = []
28      for item in items_list:
29          if top_frequency_dict[item] > 0:
30              top_frequency_items.append(item)
31              top_frequency_dict[item] -= 1
32      return top_frequency_items
```

8.4 Collision Time of Frogs

There are two frogs and each frog has a starting point and a motion vector, and it is represented by a 4-tuple as (sx, sy, dx, dy), where (sx, sy) denotes the starting point that starts from zero, and (dx, dy) that is constant direction vector of movement for each succeeding hop. Overall, each of the frogs is represented as follows: $Frog1(sx1, sy1, dx1, dy1)$ $Frog2(sx1, sy1, dx1, dy1)$. This challenge is aimed to obtain the time when both frogs jump to the same square. Write a function that as input, takes the 4-tuple for each frog and the time when both frogs jump to the same square, and if the two frogs never jump into the same square at the same time, None is returned. For some inputs, the expected outputs are illustrated in Table 8.4.

8.4 Collision Time of Frogs

Table 8.4 The expected outputs for certain inputs for finding collision time of frogs

Frog1, frog2	Expected output
(562, −276, −10, 5), (49, −333, −1, 6)	57
(−3525, −877, 4, 1), (−4405, 2643, 5, −3)	880
(2726, −3200, −6, 7), (−2290, 2272, 5, −5)	456
(1591, −1442, −10, 9), (−329, −962, 2, 6)	160

Algorithm

It takes two arguments, frog1 and frog2, representing the positions and direction vectors of two frogs. Based on the direction of the frogs, the algorithm applies certain if rules to determine the time at which the two frogs will collide.

Code 8.5 Python code to compute and return the collision time of the given frogs

```
1   def Collision_time_of_frogs(frog1,frog2):
2       '''
3       Based on Frog1(sx1,sy1, dx1, dy1) and
4       Prog2(sx1, sy1, dx1, dy1), the positions
5       are filled.
6       '''
7       sx1=frog1[0]
8       sx2=frog2[0]
9       sy1=frog1[1]
10      sy2=frog2[1]
11      dx1=frog1[2]
12      dx2=frog2[2]
13      dy1=frog1[3]
14      dy2=frog2[3]
15
16      tx=0
17      ty=0
18      '''
19      1:
20      If the constant direction vector is equal to
21      (dx1==dx2==0 and dy1==dy2==0), it means
22      that it does not move at all:
23
24      If condition: If the starting position of
25      both frogs is the same, our output will
26      be equal to 0, that is, at time 0, these
27      two are standing in the square.
28
29      Else condition: If the starting position of
```

```
30      both frogs is not the same, the output
31      will be None.
32      '''
33      if dx1==dx2==0 and dy1==dy2==0:
34          if sx1==sx2 and sy1==sy2:
35              t=0
36              return t
37          else:
38              t=None
39              return t
40
41      '''
42      2:
43      First if condition: if the constant direction
44      vector is on the y-axis, and (dx1==dx2==0),
45      The computations for the y position are done:
46
47      Second if condition: If ty is a non-decimal number
48      and ty is a positive number, ty is returned.
49
50      Else condition: If the above conditions are not met,
51      None is returned.
52      '''
53      if dx1==dx2==0:
54          ty=(sy2-sy1)/(dy1-dy2)
55          if int(ty)==ty and ty>0 :
56              t=int(ty)
57          else:
58              t=None
59              return t
60
61      '''
62      3:
63      First if condition: if the constant direction
64      vector is on the x-axis, and (dy1==dy2==0),
65      the computations for the x position are done:
66
67      Second if condition: If tx is a non-decimal
68      number and tx is a positive number,
69      tx is returned.
70
71      Else condition: If the above conditions
72      are not met, None is returned.
73      '''
74      if dy1==dy2==0:
```

8.4 Collision Time of Frogs

```
75              tx=(sx2-sx1)/(dx1-dx2)
76              if int(tx)==tx and tx>0:
77                  t=int(tx)
78              else:
79                  t=None
80              return t
81
82      '''
83      4:
84      If dx1!=dx2 is true (they are inequal):
85      1: if dy1!=dy2, tx and ty are computed.
86      1.2: If tx is equal to ty and their value
87      is non-decimal and also greater than 0,
88      tx is returned, otherwise, None is returned.
89      2: if dy1=dy2, tx is computed.
90      2.1: If tx is a non-decimal number and tx is
91      a positive number, tx is returned,
92      otherwise, None is returned.
93      '''
94      if dx1!=dx2:
95          if dy1!=dy2:
96              tx=(sx2-sx1)/(dx1-dx2)
97              ty=(sy2-sy1)/(dy1-dy2)
98              if tx==ty and int(tx)==tx and tx>0 :
99                  t=int(tx)
100             else:
101                 t=None
102                 return t
103         elif dy1==dy2:
104             tx=(sx2-sx1)/(dx1-dx2)
105             if int(tx)==tx and tx>0 and sy1==sy2:
106                 t=int(tx)
107             else:
108                 t=None
109                 return t
110
111     '''
112     5:
113     If dx1==dx2 (they are equal)
114     1: if dy1!=dy2, ty is calculated.
115     1.1: If ty is a non-decimal number
116     and ty is a positive number,
117     tx is returned, otherwise,
118     None is returned.
119     2: if dy1==dy2, None is returned.
```

```
120            ...
121        if dx1==dx2:
122            if dy1!=dy2:
123                ty=(sy2-sy1)/(dy1-dy2)
124                if int(ty)==ty and ty>0 and sx1==sx2:
125                    t=int(ty)
126                else:
127                    t=None
128                return t
129            elif dy1==dy2:
130
131                t=None
132                return t
133        return t
```

8.5 Positioning in Wythoff Array

The Wythoff array is an infinite two-dimensional grid that is characterized by its first row, which is identical to the Fibonacci sequence $(1, 2, 3, 5, \ldots)$. The array can be defined as follows: Let pr denote the previous row, cr denote the current row, and a and b denote the first and second elements of pr, respectively. The first element of the next row (excluding the first row) is determined as the smallest number that does not appear in any of the previous rows, and is denoted by c. The second element of the next row is determined based on the following two conditions:

(1) If $c - a = 2$, then the value of the second element is $b + 3$. (2) If $c - a \neq 2$, then the value of the second element is $b + 5$.

$$x = \begin{cases} b + 3 \; if \; c - a == 2 \; (equal \; with \; 2) \\ b + 5 \; if \; c - a! = 2 \; (unequal \; with \; 2) \end{cases} \quad (8.1)$$

When the first two elements are specified, the same as in the Fibonacci sequence, the next element is equal to the summation of the two previous numbers. For example, let $x1$, $x2$, and $x3$ denote the first, second, and third rows of a Wythoff array.
$x1 = 1, 2, 3, 5, 8, 13, 21, 34, 55, 89,$
$x2 = 4, 7, 11, 18, 29, 47, 76, 123, 199,$
$x3 = 6, 10, 16, 26, 42, 68, 110, 178, 288.$

Based on the preceding rows, the first element of $x2$ is 4, which is the smallest number that does not appear in $x1$. To determine the second element of $x2$, we use $a = 1$ and $b = 2$ from $x1$, and $c = 4$ from $x2$. We calculate $c - a$ which is equal to $4 - 1 = 3$. Since $3 \neq 2$, we use the formula $b + 5$ and obtain 7 as the value for the second element of $x2$. After obtaining the first two elements of $x2$, we can continue the sequence using the Fibonacci sequence. For instance, the third position is

8.5 Positioning in Wythoff Array

Table 8.5 The expected outputs for certain inputs for finding position in Wythoff array

n	Expected output
1042	(8, 8)
13	(0, 5)
127	(48,0)
20022	(4726, 1)

obtained by adding 7 and 4 from the previous two positions, resulting in 11. Using a similar approach, we can obtain $x3$ from $x2$. The objective of this challenge is to find the position (index) of a given number n as a two-tuple (row, column). To accomplish this, we need to write a function that takes a positive integer n as input and returns its position as a two-tuple (row, column). For some inputs, the expected outputs are illustrated in Table 8.5.

Algorithm

Let n be the given input. The algorithm starts by initializing the variables and sets. It then iterates over the maximum number of columns allowed ($n + 1$) and generates the two numbers in the current row based on the smaller and larger number from the previous row. The smaller number is selected as the starting point, and the algorithm iterates until an unvisited number is found. The next number based on provided rules in Eq. 8.1 is computed. The algorithm then checks if the target number is in the current row, then returns the position if found. If not found, the algorithm adds the two new numbers to the set of visited numbers, checks the rest of the row for the target number, and moves to the next row. This process repeats until the maximum number of columns is reached. The Python code to return the position n in Wythoff array is depicted in Code 8.6.

Code 8.6 Python code to return the position n in Wythoff array

```python
def Positioning_in_Wythoff_Array(n):
    # Initialize variables and sets
    visited_numbers = set([])
    current_row = 0
    smaller_number, larger_number = -1, -1
    
    '''
    Iterate over the maximum number
    of columns allowed
    '''
    for _ in range(n+1):
        
        if len(visited_numbers) == 0:
            previous_number, last_number = 1, 2
        else:
```

```python
16              '''
17              Choose the smaller number
18              as the starting point
19              '''
20              current_number = smaller_number
21
22              # Iterate until an unvisited number is found
23              while True:
24                  if current_number not in visited_numbers:
25
26                      # Calculate the next number in the row
27                      if current_number - smaller_number==2:
28                          next_number = larger_number + 3
29                      else:
30                          next_number = larger_number + 5
31
32                      '''
33                      Return the two numbers
34                      for the next row
35                      '''
36
37                      previous_number, last_number =\
38                      current_number, next_number
39                      break
40
41                  '''
42                  Increment the current number
43                  and try again
44                  '''
45                  current_number += 1
46
47              '''
48              Update the variables
49              for the next iteration
50              '''
51              smaller_number, larger_number =\
52              previous_number, last_number
53
54              '''
55              Check if the target number
56              is in the current row
57              '''
58              if n == previous_number:
59                  return (current_row, 0)
60              elif n == last_number:
```

```
61                return (current_row, 1)
62
63        '''
64        Add the two new numbers to
65        the set of visited numbers
66        '''
67        visited_numbers.add(previous_number)
68        visited_numbers.add(last_number)
69
70        '''
71        Check the rest of the row
72        for the target number
73        '''
74        column_number = 2
75        while last_number <= n:
76            new_number = previous_number + last_number
77
78            '''
79            Check if the target number
80            is in this column
81            '''
82            if new_number == n:
83                return (current_row, column_number)
84
85            # Update the variables for the next iteration
86            previous_number = last_number
87            last_number = new_number
88            visited_numbers.add(new_number)
89            column_number += 1
90
91        # Move to the next row
92        current_row += 1
```

8.6 Fractran Interpreter

Fractran is a programming language that uses fractional numbers to define its esoteric programs. The goal of this challenge is to provide an interpreter for Fractran. To interpret a given program, an integer 'n' is multiplied by the fractions at a given state until an integer is produced, which signals the end of the program. Initially, the state is equal to 'n', and in subsequent steps, it is updated. For example, suppose we are given an integer 'n' of 3, a program consisting of fractions [(7, 3), (12, 7)], and a number of iterations 'itera' set

Table 8.6 The expected outputs for certain inputs for interpreting a Fractran program

n, program, itera	Expected output
6, [(228, 131), (327, 158), (161, 394), (77, 425)], 4	[6]
2, [(345, 162), (108, 125), (90, 45), (293, 277)], 5	[2, 4, 8, 16, 32, 64]
10, [(1, 145), (7, 349), (156, 24)], 65	[10,65]

to 4. In this case, the task is to interpret the program. At first, a pointer i is to be defined, where at least $i = 0$, and at most $i = |itera| - 1$, and from $\frac{7}{3}$, 7 is multiplied by 3, and 21 is divided by 3, which 7 is stored in $temp = 7$, and $temp$ is stored into array R. In the next step, as $temp$ is an integer, $state$ is updated and is equal to 7, and from the beginning in $program$, multiplying is performed, because in the previous step, an integer is found. Hence, i is incremented by one and i is to be updated from zero to one. From $\frac{7}{3}$, $temp = 7 \times 7/3 = 16.33$, so the temp is a fractional number, and (12, 7) (the next fraction) in $program$ is considered, and from $\frac{12}{7}$, $temp = 12 \times 7/7 = 12$. Now, $temp$ is an integer, so $temp$ is stored in R. Again, from the beginning in $program$, multiplying is performed and $state = 12$, hence, i is incremented by one and i is to be updated from one to two. From $\frac{7}{3}$, $temp = 7 \times 12/3 = 28$, so the temp is integer number that is stored into R, and $state = 28$, and $i = 3$. From beginning in $program$, multiplying is performed, where from $\frac{7}{3}$, $temp = 7 \times 28/3 = 65.33$. So, the temp is a fractional number, and (12, 7) in $program$ is considered, and from $\frac{12}{7}$, $temp = 12 \times 28/7 = 48$. Now $i = |itera| - 1$, it means all iterations are done, and R is the interpretation made. Write a function that as input, takes a positive integer n and an array of fractions, and the number of iterations, and as output, returns an array of the integers produced by the Fractran Program. For some inputs, the expected outputs are illustrated in Table 8.6.

Algorithm

The Algorithm to interpret a Fractran program uses fractions to generate a sequence of numbers. Given an initial number and a set of fractions, the algorithm multiplies the initial number by each fraction (in turn) and then checks whether the result can be simplified to an integer. If it can, the integer is added to the sequence and becomes the new input number for the next iteration. The process continues until a certain number of iterations has been reached, or until no more integers can be generated. With details, the algorithm has the following steps:

1. It takes in three parameters: n, $program$, and $itera$, where n is the initial value to start with, $program$ is a list of fractions represented as tuples (numerator, denominator) and $itera$ is the number of iterations to run the program for
2. It then initializes a list called $result$ with the initial value n, and sets the numerator and denominator to n and 1, respectively.

8.6 Fractran Interpreter

3. Next, it runs a loop for *itera* number of times, and within that loop, it runs another loop to iterate over each fraction in the program list.
4. For each fraction, the algorithm multiplies the numerator by the first number in the fraction, and multiplies the denominator by the second number in the fraction.
5. It then calculates the greatest common divisor (GCD) of the resulting numerator and denominator using the gcd, and reduces the numerator and denominator by dividing both by the GCD (GCD is used to simplify the fractions).
6. If the resulting denominator equals 1, the algorithm adds the numerator to the *result* list (it is visited an integer number), updates the numerator and denominator to the reduced values, and breaks out of the inner loop.
7. After the outer loop finishes running for the specified number of iterations, the algorithm returns the list of generated values.

The Python code to interpret a Fractran program is depicted in Code 8.7.

Code 8.7 Python code to interpret a Fractran program

```
'''
This function calculates the greatest
common divisor (GCD) of two numbers
'''
'''It takes two arguments a and b,
 and returns their GCD.
'''
def gcd(a, b):
    if b == 0:
        return a
    else:
        return gcd(b, a % b)


'''It takes three arguments: n -
 the initial value, program - a list of fractions,
 and itera - the number of iterations to run.
'''
def Fractran_Interpreter(n, program, itera):
    # Initialize the list with the initial value
    result = [n]
    numerator = n
    denominator = 1
    for i in range(itera):
        for j in range(len(program)):
            '''
            # Multiply the numerator by the first
            number of the fraction
            '''
            temp_num = program[j][0] * numerator
```

```
                    '''
                    # Multiply the denominator by the
                    second number of the fraction
                    '''
                    temp_denom = program[j][1] * denominator
                    '''
                    # Calculate the GCD of the numerator
                    and denominator
                    '''
                    div = gcd(temp_num, temp_denom)
                    '''
                    Reduce the numerator by dividing
                    by the GCD
                    '''
                    temp_num //= div
                    '''
                    # Reduce the denominator
                    by dividing by the GCD
                    '''
                    temp_denom //= div
                    '''
                    # If the denominator is 1, add
                    the numerator to the result and
                    update the numerator and denominator
                    '''
                    if temp_denom == 1:
                        result.append(temp_num)
                        numerator = temp_num
                        denominator = temp_denom
                        '''Stop checking fractions once one
                        is found that can be applied
                        '''
                        break
        '''
        # Return the list of values
        generated by the program
        '''
        return result
```